D1174971

Boundary Layers
of Flow and Temperature

Boundary Layers of Flow and Temperature

BY ALFRED WALZ

Edited and Translated from the German
by Hans Joerg Oser

The M.I.T. Press
CAMBRIDGE, MASSACHUSETTS, AND LONDON, ENGLAND

LIBRARY
FLORIDA STATE UNIVERSITY
TALLAHASSEE, FLORIDA

Originally published in German by Verlag G. Braun in Karlsruhe
under the title "Strömungs- und Temperaturgrenzschichten"

English translation Copyright © 1969 by
The Massachusetts Institute of Technology

Designed by Dwight E. Agner. Set in Monotype Times Roman and
printed and bound in the United States of America
by The Riverside Press, Inc.

All rights reserved. No part of this book may be reproduced in
any form or by any means, electronic or mechanical, including
photocopying, recording, or by any information storage and retrieval
system, without permission in writing from the publisher.

SBN 262 23035 6

Library of Congress catalog card number: 69–12761

LIBRARY
FLORIDA STATE UNIVERSITY
TALLAHASSEE, FLORIDA

To Henry Görtler

this volume is dedicated jointly
by the author and the translator

Preface to the German Edition

It is customary to develop boundary layer theory from the Navier-Stokes theory of viscous flow for the special case of large Reynolds numbers. In that process, *those* terms in the Navier-Stokes equations are neglected which are generally of secondary physical importance and, moreover, mathematically unpleasant.

The deliberate limitation to the most interesting case of large Reynolds numbers makes it possible to derive Prandtl's boundary layer equation and the energy equation for the temperature boundary layer directly, based on the same physical arguments that were used to justify neglecting the before-mentioned terms in the Navier-Stokes equations and in the general differential equation for the temperature field. Doing so makes it possible to present an easily comprehensible, from the mathematical and physical point of view, introduction into the theory of flow and temperature boundary layers, a topic usually considered to be in the domain of higher mathematics and avoided by students in the pure engineering disciplines, which becomes interesting to practicing engineers only when absolutely necessary, i.e., when the classical theory of inviscid flow, or hydraulics, as it is called, fails patently. One of the aims of this book is to erase the impression that boundary layer theory can be understood and successfully applied by specialists only.

From an understanding of the simplified assumptions that form the foundation of boundary layer theory, physically and mathematically well charted paths lead on the one hand to a general foundation of Navier and Stokes' theory, and on the other to the applications of the theory. After formulating the simplified theoretical foundations for laminar and turbulent boundary layers in incompressible and compressible flow without and with heat transfer, this book pursues only the second path, the *applications of the theory*.

A good deal of space in this book is devoted to a presentation of the method of integral conditions with the implied intent to use it later in a boundary layer theory of second order which will be capable of accounting for the interaction between boundary layer and potential flow, a desirable problem to be solved from a modern point of view. Based on new mathematical methods of error estimating (see the papers by K. Nickel [80, 81, 82]) and on many exact comparisons in difficult problems, the suspicion has been removed that this way of approximating boundary layer solutions leads to good results merely by sheer accident. The power of the method is rather based on good physical and mathematical evidence. In this context we refer to a publication by A. A. Dorodnitsyn [14].

The offering of a collection of quadrature formulas, nomograms, and computational schemes for the purpose of facilitating practical computations with slide rules and desk calculators would appear at first glance as a lost effort. Practical work teaches us, however, that for purposes of getting oriented about the orders of magnitude of boundary layer phenomena and for small problems, the computation "by hand" may still be the faster, cheaper, and, above all, the didactically more valuable way. For systematic boundary layer investigations the use of an electronic computer is, of course, indicated. The quadrature formulas and the analytic representations developed in this book represent a suitable starting point for a computer program.

It seemed necessary to the author, for didactical reasons, to point out the common features of the theories for laminar and turbulent boundary layers and to present, as far as possible, the equations in a form that holds for boundary layer states.

In the age of high-velocity techniques, of supersonic flight of aircraft and rockets in the atmosphere, boundary layer theory had to become broader insofar as it had to deal not only with *flow boundary layers* but with *temperature boundary layers* as well. On one hand, because these boundary layers influence one another, to make it possible to compute the one, the other has to be accounted for; on the other hand, the science of heat transfer can be lifted from its rather empirical state only with the aid of boundary layer theory.

The temperature boundary layer and problems in heat transfer are therefore treated here with the same attention to detail as are flow boundary layers. Within the framework of the approximation theory with integral conditions, which is to be presented here, it is fortunately possible, without undue mathematical difficulties, to include in the differential equations for the flow boundary layer the very general case of variable material properties,* i.e., for example, the case of compressible laminar

*The specific heat for constant pressure, c_p, is assumed constant, within the framework of the approximation theory presented here.

and turbulent flow with heat transfer.

For the engineer, theory becomes alive and interesting only with a practical example. The theoretical part of this book is therefore augmented by a section containing numerous examples that are arranged in increasing degree of difficulty. Problems more fundamental in nature, which occur often in the context of boundary layer problems, such as the laminar-turbulent transition problem, the determination of the maximal lift of airfoils, influencing the boundary layer by suction, and so on, are treated briefly in Chapter 5.

Like every other author who reports on results of a rapidly expanding branch of research, this author is also faced with the question of conscience whether the latest and most useful state of the art is presented. It is the custom to forgive an author for overlooking an important result, perhaps even in his own specialty, in today's flood of publications or for favoring his own results or ideas over those of other workers, because of a lack of time for the concentrated study of those other publications. The author of this book will have to ask for indulgence of this kind for himself.

In laying out the plan of this book, the author was able to use results of research done by his colleagues (mostly Ph.D. theses at the Technical University Karlsruhe). Sincere thanks are particularly due to Drs. K. O. Felsch, H. Fernholz, D. Geropp, M. Mayer, and Mr. H. Siekmann for their help in shaping Chapter 6 and Appendix I. The author is particularly grateful for the careful and critical review of the manuscript by Dr. W. Schönauer, and to Miss R. Ulmrich for her conscientious typing and drafting. . . .

Karlsruhe, May 1966.   Alfred Walz

Translator's Preface

This translation of the German edition, which appeared in 1966 under the title "Strömungs- und Temperaturgrenzschichten," incorporates many small changes suggested by the author. Section 3.5.3.3 has been extended, and the subsequent section is completely rewritten. A new Appendix III contains flow diagrams describing the three computational methods developed in the text. Inevitably some of these methods would be different had they been developed for an electronic computer. But the original intent — and some of the computational techniques go back to the early forties — was to aid the aerodynamical engineer by providing him with a tool that would allow him to compute by hand, quickly and reasonably precise, the important parameters that determine the boundary layer flow around a given contour.

This book fills a gap in the literature. While the laminar boundary is theoretically quite well understood, there is still room for theoretical speculation in turbulent boundary layers. The author's own efforts and those of his former students have contributed considerably to a systematic treatment of both laminar and turbulent boundary layers. The translator is proud of having had the good fortune of working with the author during 1953–1957 on some of the earlier attempts of treating turbulent boundary layers computationally.

Although the editor-translator is a full-time staff member at the National Bureau of Standards in Washington, D. C., this translation was done on his own time. No facilities of the National Bureau of Standards were used during the course of this work.

Sincere thanks are due to the author who followed the birth of the English edition with patience and interest. Numerous changes were made during our correspondence, a process which is difficult at best and which can be appreciated only by two authors who have experienced writing a book together while separated by the Atlantic Ocean. Mrs. Lois Adams deserves credit for typing a flawless copy of the manuscript, and Mrs. Catherine Hartsfield for the figure captions and the index.

Washington, D. C.   Hans J. Oser
January, 1969

Contents

Summary of Notations

1. Coordinates and lengths

x coordinate along the contour of flow

y coordinate perpendicular to the contour

Y transformed y coordinate, eq. (4.71)

η nondimensional y coordinate, eq. (3.80)

δ boundary layer thickness

δ_l thickness of the laminar sublayer

δ_1 displacement thickness

δ_2 momentum-loss thickness

δ_3 energy-loss thickness

δ_4 density-loss thickness

L Length of flat plate

l_t mixing-path length

Λ wavelength of sinusoidal perturbation

$Z = \delta_2 R_{\delta_2}^n$ thickness parameter

$\Re$ cross-sectional radius of bodies of revolution

2. Velocities

u velocity in x direction

v velocity in y direction

a local velocity of sound

a^* critical velocity of sound

u_τ shear-stress velocity, eq. (3.107)

3. Other physical quantities

p static pressure

$q_\infty = \rho_\infty u_\infty^2 / 2$ stagnation pressure

ρ density

T temperature in degrees Kelvin

T_e recovery temperature, eq. (2.136)

q heat energy per unit of width and unit of length
τ shear stress
τ_s apparent shear stress in turbulent flow
$\tau_e = \tau + \tau_s$ effective shear stress in turbulent flow
ν kinematic viscosity
μ dynamic (molecular) viscosity
μ_s apparent viscosity in turbulent flow
$\mu_e = \mu + \mu_s$ effective viscosity in turbulent flow
λ molecular heat conductivity
λ_s apparent heat conductivity in turbulent flow
$\lambda_e = \lambda + \lambda_s$ effective heat conductivity in turbulent flow
c_p specific heat at constant pressure
c_v specific heat at constant volume
κ adiabatic exponent
R gas constant
β^* wedge angle

4. Nondimensional quantities

$M_\delta = u_\delta/a$ local Mach number
$M^* = u_\delta/a^*$ critical Mach number
$R_L = \rho_\infty u_\infty L/\mu_\infty$ Reynolds number of the unperturbed flow
$R_{\delta_2} = \rho_\infty u_\delta \, \delta_2/\mu_w$ local Reynolds number
Pr Prandtl number
c_f local coefficient of friction
C_F Drag coefficient
c_w profile drag
c_a coefficient of lift
$H = (\delta_3)_u/(\delta_2)_u$ shape parameter of velocity profile
$H^* = \delta_3/\delta_2$
$\Gamma = (\delta_2)_u^2/\nu \cdot du_\delta/dx$ shape parameter of velocity profile
$H_{12} = (\delta_1)_u/(\delta_2)_u$
$\Theta = (T_e - T_w)/(T_e - T_\delta)$ heat transfer parameter
K correction function for the temperature profile
c_D dissipation integral, eq. (3.139)
r recovery factor

5. Subscripts

∞ quantities in unperturbed flow
δ quantities at the edge of the boundary layer
w quantities at the fixed contour, or wall
u quantities depending *only* on the velocity distribution in the boundary layer

1   quantities referring to laminar flow
t   quantities referring to turbulent flow
J   quantities in the momentum law (4.80)
E   quantities in the energy law (4.81)
0   quantities at rest

6. Relational symbols
≈   approximately equal
~   proportional to

Boundary Layers
of Flow and Temperature

1 Introduction

1.1 Historical Development and Aim of Approximation Theories

In hydrodynamics, we ordinarily deal with water or air as the flowing medium. These media have the peculiar property to show only a minor dependence upon viscosity over a very wide range of flow parameters. It is therefore understandable that hydromechanics, already at its inception by D. Bernoulli* and L. Euler in the 18th century, was essentially confined to the limiting case of media with vanishing viscosity, the so-called "ideal fluids." This is how "classical hydrodynamics" began. Its laws are, even in the era of highly developed flow machines and fast-flying airplanes, still the basis from which the aerodynamical engineer can calculate in advance and study important properties of flows.

But it is just the aerodynamical engineer who always hits upon the limits of validity of *classical hydrodynamics:* it cannot explain the phenomena of drag, of pressure loss in pipes, or of flow separation, all of which are direct or indirect effects of the viscosity of the flowing media. The engineers of the 18th and 19th centuries knew how to help themselves by accounting for the viscosity effects through empirical laws. They developed the engineering science of *hydraulics* as a practical extension of the classical hydrodynamics of ideal fluids.

Not that there were no efforts being made to calculate theoretically the influence of viscosity on the motion of fluids. As early as the middle of the last century, Navier and Stokes [77, 127] succeeded in formulating the conditions for the equilibrium of forces (viscosity forces, inertial forces, pressure and body forces†) mathematically. The result was a system of a partial nonlinear differential equations of second order. However,

---

*We point out that the so-called Bernoulli equation was established in 1738, prior to the Euler equations, 1755.
†For example, buoyancy and centrifugal forces.

extraordinary mathematical difficulties prevented any useful application of these equations for half a century.

It was in 1904 that Prandtl [94] decisively influenced the theory of viscous flows by his proposal to simplify the Navier-Stokes equations. He showed that viscosity, as long as it is small (more precisely: for large Reynolds numbers), affects the flow only in a relatively thin "boundary layer" in the vicinity of solid walls. In this boundary layer, the velocity rises from zero at the wall (no slip) to practically the full value of the unperturbed external velocity. As a result of the relatively small thickness of this boundary layer, some of the mathematically particularly unpleasant terms in the Navier-Stokes equations become negligibly small. The result is Prandtl's boundary layer equation. It differs from Euler's equations of motion [24] for ideal fluids by only one additional term which contains the influence of viscosity.

Prandtl's boundary layer equation has become the starting point of a special area of research within hydromechanics which bridges the gap between classical hydrodynamics and the empirical science of hydraulics. A general solution of Prandtl's boundary layer equation is not yet possible, for mathematical reasons. However, it turned out very early that a solution is possible in a few special cases of practical importance. It was a major event when Prandtl's collaborator Blasius [4] was able in 1907 to compute the drag of a flat plate in laminar boundary layer flow from the solution for a special case of a flow without pressure gradient, in excellent agreement with very careful measurements.

A particular success of the theory was later the establishment of conditions for the external flow under which velocity profiles of the laminar boundary layer are similar to one another at every point. The most important contributions to the solution of this problem have been made by Falkner and Skan [25], Hartree [43], Goldstein [37], and Mangler [68]. It turned out that such "similar solutions" of the boundary layer equation are possible for the entire region of accelerated and retarded flows, whenever the external velocity is proportional to a power of the distance from the leading edge. Blasius' solution [4] for the laminar flow along a flat plate (flow without pressure gradient) turned out to be merely a special case of this class of general solutions.

The study of these similar solutions, which can be applied with somewhat different conditions to turbulent boundary layers as well, has afforded important insights in the principal properties of boundary layers in accelerated and retarded flows. By this, and in part already by the exact solution for the flat plate, the ground is laid for the development of approximation theories, since it now became possible to assume as known certain general properties of the velocity distribution within the boundary layer, the so-called "velocity profile." The problem thus having been simplified, it remained for the approximation theory merely to explain a few unknown properties of the boundary layer for arbitrarily given external velocities.

The principal idea of approximation theories may be characterized more precisely as follows: one develops a table of velocity profiles obtained from special solutions, for example, the above-mentioned "similar solutions" for increasing and decreasing pressure, or from measurements under different conditions for the external flow, and then one tries to classify these velocity profiles more or less approximately by a "shape parameter." This shape parameter of the velocity profile may be introduced as an unknown variable into the approximation theory. A second unknown could suitably be the thickness of the boundary layer. Mathematically, this means that one introduces a trial solution for the velocity profile which possesses the properties of the special "similar solutions" or experimental velocity profiles and which contains at least two free coefficients (a shape parameter and the boundary layer thickness) that allow fitting the solution to the general problem (boundary conditions).

An approximation theory of this type is expected to determine the two (or more) unknown coefficients for a given *arbitrary* external velocity distribution along a fixed wall. Just as in the exact formulation of the boundary layer problem, we have available the laws of conservation of mass, momentum, and energy as well as certain boundary conditions. However, since in the simplified problem of the approximation theory the dependence of the velocity profile upon the distance from the wall is assumed to be known except for the influence of a shape parameter and the boundary layer thickness itself, these conservation laws result in *ordinary differential equations* in the form of integral conditions, in place of the partial differential equations of the exact theory. From these equations, the unknowns introduced into the theory (for example, the shape parameter and the boundary layer thickness) may be determined by elementary mathematical methods related to the principles of variational calculus.

A method for calculating the incompressible *laminar* boundary layer according to this basic idea was first proposed by von Kármán [56] and Pohlhausen [91] in 1921. The system of equations for the two unknowns of the approximation theory consisted in a condition for the balance of forces averaged over the boundary layer thickness, the "momentum integral condition," and a boundary condition for the balance of forces that act upon a particle in the immediate vicinity of the wall (where velocity and inertial forces vanish). This boundary condition, which may serve as the second equation of the approximation theory, is referred to in the boundary layer literature as "compatibility condition." The velocity profiles of the laminar boundary layer were approximated by polynomials of fourth degree in a variable which is the ratio of the distance from the wall and the boundary layer thickness; the coefficients of this polynomial depend only on the shape parameter of the velocity profile.

A similar method of computation for the incompressible *turbulent boundary layer* with an empirically determined one-parameter "catalog" of velocity profiles, together with an integral condition for the momentum and the no-slip condition as a system of equations was given by Buri [5] in 1931. Gruschwitz [41] in the same year improved the semiempirical approach by which Buri computed turbulent boundary layers by adding to the momentum integral condition (with empirical estimates about shear stress) a kind of integral condition for the energy, using empirical constants derived from dimensional analytical considerations. He used there, for the first time, the "momentum-loss thickness" as a characteristic quantity whose importance in simplifying the computation of laminar boundary layer was, strangely enough, not recognized until ten years later by Holstein and Bohlen [47] as well as (independently), at the same time, by Tani [132] and Koschmieder and Walz [59].

A milestone in the development of approximation theories for laminar and turbulent boundary layers turned out to be the paper of K. Wieghardt [155], written in 1944 (but not published until 1948), in which he used an integral condition for the energy resulting from partial integration of Prandtl's boundary layer equation.* Thus, there became available a system of three equations — the integral conditions for momentum and energy and a compatibility condition — usable as a starting point for approximation theories. Based on important general investigations about polynomial approximations by Mangler [69], Wieghardt [155] was able to improve the trial solutions for the laminar boundary layer by introducing two shape parameters for the velocity profile in addition to the boundary layer thickness. It was thus possible to improve the accuracy of the approximate solutions for retarded flows quite appreciably. The method failed, however, in accelerated flows. Moreover, the amount of labor was so much larger in comparison to the one-parameter method that the method did not take hold.

In 1944 Walz [140] was able to prove that the one-parameter methods result in far better answers when integral conditions for momentum and energy are used and the compatibility condition is disregarded. Most practical boundary layer computations have been carried out on the basis of such an approximation theory up until the present time.

---

*It became known to the author only recently that Leibenson [65] derived an integral condition for the energy as early as 1935.

For *turbulent boundary layers*, the integral conditions for momentum and energy in the general form given by Wieghardt* were not useful until reliable empirical laws for shear stress and dissipation in turbulent boundary layers became available. Among the numerous papers that have been written to justify these empirical laws (at first only for incompressible flows), not only those by von Kármán [54], Schoenherr [115], and Ludwieg and Tillmann [67] concerning sheer stress should be mentioned, but also those by Rotta [108] and Felsch [27] on the dissipation law, which were particularly fruitful.

Because of the increasing importance of compressible flows with or without heat transfer, it became important to extend both exact boundary layer theory and approximation theory to this general case. The theory becomes more complicated in this case because the density is now a function of pressure and temperature, while viscosity depends essentially on temperature only, but also because the kinetic energy and frictional work in the boundary layer flow reach the order of magnitude of the enthalpy and for that reason changes in velocity are followed by changes in temperature. When formulating the conservation laws for mass, momentum, and energy, the laws of thermodynamics and gas dynamics have to be observed. This means that along with the *flow boundary layer*, there are also the *temperature boundary layer* and the mutual influence of these boundary layers upon one another to be accounted for.

It is a very fortunate circumstance that the partial differential equation for the temperature boundary layer is linear and therefore offers ideal conditions for finding exact solutions. Busemann [6] was, already in 1935, able to show in the important case of gas flows that the temperature field is uniquely coupled with the velocity field under certain assumptions that are quite often met exactly in both laminar and turbulent boundary layers.

The *approximation theory of compressible boundary layers* allows therefore a rather far-reaching simplification: the catalog of velocity profiles that may be one- or multiparametric is coupled with a table of temperature profiles that contain no additional unknowns of the approximation theory. The Mach number of the free-stream flow is used as a measure of compressibility, and the parameter for heat transfer appears also in the equation that couples temperature and velocity profiles, but both may be considered as known functions of the distance from the leading edge of the boundary layer for any problem.

---

*Gruschwitz [41] derived his empirical energy equation from a limited number of measurements and was, therefore unable to prove its general validity. Von Doenhoff and Tetervin [12] tried to overcome this disadvantage by using a very large number of measurements. However, only the integral condition for energy as given by Wieghardt in the universal form appears physically plausible.

In *compressible turbulent boundary* layers with and without heat transfer, it is not sufficient to know the connection between velocity and temperature boundary layers in order to compute practically with the integral conditions for momentum and energy. One needs, moreover, the empirical laws for shear stress and dissipation in compressible flows with heat transfer. For a number of years now there have been known results from difficult experimental and semiempirical research which are not completely understood in all cases but which may still provide a base for an approximation theory sufficiently accurate as a rule.

In this survey on the development of approximation theories, we should mention a later result, which derives from the paper by Wieghardt [155] on the computation of laminar boundary layers in incompressible flows. Through a suitable generalization of the polynomial approximation for the velocity profile, Geropp [33] was able to overcome the difficulties in Wieghardt's method for accelerated flow and at the same time to include the case of compressible flows with and without heat transfer by means of a suitable transformation of coordinates. The numerical effort required for computing a practical example is surprisingly small.

By using a two-parameter family for the laminar velocity profiles, such a good approximation of known exact solutions can be achieved that for purposes of the practical engineer further refinement of the approximation theory does not appear promising, in general. However, it is possible, in principle, to refine the method by allowing for additional integral conditions and shape parameters for the velocity profiles.

But this statement does not eliminate the question how exact the approximate solutions are in cases hitherto incapable of being checked. Until recently it seemed unclear whether generally valid statements could be made about the accuracy of approximation solutions based on Prandtl's boundary layer equations. The papers by Nickel [80, 81] on incompressible laminar boundary layers have substantially increased the understanding insofar as it is now possible to develop generally valid upper and lower bounds for the exact solution. A generalization of these investigations for compressible flow with and without heat transfer has not yet been possible, but is feasible in principle. We should mention here the rather favorable experience with the method of integral conditions reported by Dorodnitsyn [14]. Until turbulent phenomena can be treated theoretically in a rational way, we must depend on experiments for checking the validity of the approximate solutions for turbulent boundary layers.

In the case of the laminar boundary layer, the question arises whether it is not possible to improve the approximate solutions by iteration based on the exact boundary layer equations. This path was chosen by Kwang-

Tzu Yang — first for "similar" solutions [61] and later for more general laminar boundary layers [62]. His few successful examples do not allow to decide whether this way to check approximate solutions, although in principle ideal, is generally feasible.

Other methods, which proved successful many times in finding exact solutions for Prandtl's boundary layer equation and the equation for the temperature boundary layer, are the so-called "finite-difference methods," and also the method of series expansions. The former consist in replacing the partial differential equations by difference equations and then computing the velocity and temperature profiles in a step-wise fashion from the previous ones. Papers dealing with the difficulties of convergence and numerical stability that occur with these methods — particularly in adverse pressure regions and in the vicinity of the separation point — have been published by Görtler [38, 39], Witting [158], and more recently, with particular success, Schönauer [116]; (additional references are listed in [116]). Both the finite-difference and series expansion methods require electronic computers to carry out the calculations in an acceptable amount of time.

Whenever in the application of computational methods, such as that by Schönauer, the proof of numerical stability is possible, the results may be considered exact solutions of Prandtl's boundary layer equations. More recently Wippermann [157] developed these methods further to include compressible laminar boundary layers as well.* It is therefore possible to state that the problem of the laminar boundary layer is solved in principle today. However, approximation theories and methods that can be checked by exact solutions will remain important for the engineer, even in the case of laminar boundary layers.

Any presentation of an approximation theory would be incomplete if it would not show explicitly which problems can be solved with it in practice, how the computations should be carried out, and what degree of agreement is possible with available exact solutions or measurements. Ample space is devoted to achieve this goal in this book. It may already be stated here in the Introduction that the numerous examples will demonstrate the usefulness of the developed approximation theory — even where the velocity profiles are characterized by a single shape parameter.

It should also be noted here that this book, by pointing to the extensive and thorough presentation of the material in the excellent textbook by Schlichting [111], may limit itself to listing only the most important

---

*D. B. Spalding [124] showed that finite-difference methods may be applied in the case of turbulent boundary layers, provided the apparent viscosity is assumed to be known.

relations of exact boundary layer theory; this is the starting point and basis of the approximation theory, which is the main subject. Didactical reasons make it appear justifiable to limit the considerations in the present volume to stationary boundary layer phenomena and thus, with the exception of axisymmetric flows, to consider only the plane problem. Concerning the problem of laminar-turbulent transition, we may essentially refer to the extensive coverage in Schlichting's book, so that it may suffice here to present only the most important results of that theory (Section 5.1). Once the reader masters this particular material, it should not be difficult for him to understand also the more general formulations and ways of solution in boundary layer theory that are omitted here; they can be found in the book by Schlichting or in the references to a rapidly increasing literature in this field, as listed by him.

In the treatment of temperature boundary layers, for the same didactical reasons only forced convection is dealt with, a case which appears most often in practical applications, and we restrict ourselves almost exclusively to the flow of gases that are assumed ideal and have a Prandtl number of approximately 1. Despite these limitations, that section occupies a good part of this book, since more recent results of the exact theory had to be included which are less well known but which form the basis for approximate solutions as well as for current problems in heat transition.

In Chapter 5, the following additional material pertaining to the approximation theory is treated briefly: computation of the laminar-turbulent transition, axisymmetric boundary layers, heat transfer in strong temperature and pressure gradients, computation of potential flow about profiles allowing for interaction with the boundary layer, computation of the maximal lift and drag of airfoils, and a simple theory of boundary layer suction by single slots.

The serious reader of this book should be able to apply the developed theory immediately, that is to say without extensive preparations for computational techniques. In Appendix I all essential equations are therefore listed together with explicit instructions for computation. Appendix II finally contains some gas-dynamical relations (with references to the appropriate literature) which are needed in the theory of compressible boundary layers.

1.2 Physical Interpretation of Viscosity

Viscosity forces play an important role in boundary layer theory. For this reason it is useful to start out with some remarks about the physical phenomenon of viscosity.

When speaking of viscosity in liquid and gaseous media, one refers to the phenomenon that the relative displacement of neighboring particles requires a certain force. Such a relative displacement of particles takes place, for example, in laminar parallel flow where the velocity changes from layer to layer. The force that occurs between neighboring layers (or between the particles in these layers) per unit of area is called "shear stress." Measurements indicate that this shear stress is proportional to the velocity gradient perpendicular to the direction of flow with a proportionality factor that is a material property called the *viscosity*. Denoting by u the velocity component in x direction (Fig. 1.1), the viscosity by

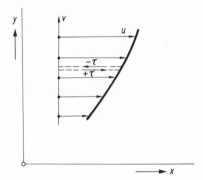

Fig. 1.1. Illustration of Newton's law of shear stresses.

μ and the shear stress by τ, the observed phenomenon may, according to Newton [79], be described by the simple law:

$$\tau = \mu \frac{\partial u}{\partial y}. \tag{1.1}$$

Here τ is a force per unit area which acts in positive or negative x direction, depending on which of the two interacting layers is being considered. For a layer that flows more rapidly than its vicinity, τ is a retarding force and thus acts in a negative direction. The neighboring, more slowly moving, layer experiences an accelerating force in the direction of flow.

How can the origin of these shearing forces be explained physically? The theory of molecular motion answers this question completely. A distinction has to be made between gaseous and liquid media.

In gases, the individual molecules undergo random (3-dimensional) motions. After traveling a certain length, called the *free mean path*, they collide with other molecules. An exchange of momentum takes place during that collision. The free mean path in air under normal atmospheric conditions (pressure 1 atm, temperature 20° C) is of the order of 10^{-7} m. The velocity of the molecules reaches extremely high peak values which are of the order of 10^3 m/sec. When a microscopic flow is superimposed on the medium (that is, to the molecules as a whole), for example in x direction, then the molecules assume an average velocity u upon which the random molecular velocity is superimposed. At any given time a certain number of molecules move perpendicularly to the direction of flow x (that is, in y direction). If there exists in y direction a velocity gradient $\partial u/\partial y$, these molecules not only carry mass but also momentum into the more slowly or more rapidly moving neighboring layer.

The mass m which enters the neighboring layer from the original one, per unit of area and time, is proportional to the density and the average molecular velocity $\mathring{v}$ in y direction. We have therefore

$$m = \rho\mathring{v} \tag{1.2}$$

(mass per unit of area and time). The mass m had the velocity u_1 in the original layer. After moving the length of the mean-free path l in y direction, this mass has a new velocity because of its penetrating into this neighboring layer:

$$u_2 = u_1 \pm l\frac{\partial u}{\partial y}. \tag{1.3}$$

The mass m has therefore changed its velocity by the amount:

$$u_2 - u_1 = \pm l\frac{\partial u}{\partial y}. \tag{1.4}$$

This change of velocity corresponds to a change of momentum which is equivalent to a force acting upon this neighboring layer. The *shear stress* τ as a force per unit area from eqs. (1.2) and (1.4) is equal to

$$\tau = -m(u_2 - u_1) = -\rho\mathring{v}l\frac{\partial u}{\partial y}. \tag{1.5}$$

τ is positive (in x direction) when v is negative and $\partial u/\partial y$ is positive. For this reason we have put a negative sign in eq. (1.5).

The theory of molecular motion tells further that the mean-free path l and the density ρ are inversely proportional:

$$l \sim \frac{1}{\rho}. \tag{1.6}$$

The shear stress is therefore independent of the density ρ. The factor remaining with $\partial u/\partial y$ is a material property, which increases with temperature because of the fact that the molecular velocity $\mathring{v}$ increases with temperature. This quantity, which is proportional to v, is nothing but the *viscosity* μ of the medium which was introduced already in eq. (1.1). For *gases*, the following empirical relation holds, according to Sutherland [131], with very good accuracy:

$$\mu(T) = C_1 \frac{T^{3/2}}{T + C_2}. \tag{1.7}$$

Viscosity in units of N sec/m^2 (N = newton), temperature in $^\circ$K with $C_1 = 1.486 \times 10^{-6}$ and $C_2 = 110.6$, for air.*

In *fluids* the origin of viscosity is different, though again of molecular nature. The molecules in fluids are relatively close to another, such that there is no room for a large random motion as observed in gases. However, the molecules are not yet packed as densely as in solids, where the attempt of deformation of the medium is met by shear stresses resulting from elastic forces. In fluids, the displacement of individual molecules is still possible when they are subjected to shearing forces. Velocity differences between neighboring layers result in a type of sliding motion (no longer associated with exchange of momentum). The shear stresses that occur between these layers quite often obey also Newton's law (1.1). The viscosity μ for fluids is also a quantity that depends on the medium. However, contrary to the viscosity of gases, it decreases with increasing temperature.

According to [139], we have for water in the range 283° K $< T < 383^\circ$ K with $\pm 1\%$ accuracy:

$$\mu(T) = 1.71586 \times 10^{-3} - 4.83556 \times 10^{-5}(T - 273)$$
$$+ 7.58441 \times 10^{-7}(T - 273)^2 - 6.19683 \times 10^{-9}(T - 273)^3$$
$$+ 2.01591 \times 10^{-11}(T - 273)^4 \quad \text{in N sec/m}^2, \tag{1.10}$$

and for oil, according to [73] in the range 283° K $< T < 383^\circ$ K, again with $\pm 1\%$ accuracy:

$$\mu(T) = 3.10773 \times 10^{-2} - 1.04622 \times 10^{-3}(T - 273)$$
$$+ 1.28857 \times 10^{-5}(T - 273)^2 - 5.44828 \times 10^{-8}(T - 273)^3$$
$$\text{in N sec/m}^2. \tag{1.11}$$

*Mathematical reasons account for often writing eq. (1.7) in the approximate form

$$\frac{\mu}{\mu_1} \approx \left(\frac{T}{T_1}\right)^\omega, \tag{1.8}$$

with

$$\omega = \tfrac{3}{2} - 1/(1 + C_2/T_1) \tag{1.9}$$

and $0.5 < \omega < 1.5$ ($\omega \approx 0.8$ for $200 < T_1 < 400^\circ$ K).

Additional relations for these and other gases and fluids can be found,
for example, in [139]. Fluids whose viscous behavior cannot be described
by Newton's law (1.1) shall not be considered within the framework of
this book

1.3 "Apparent" Viscosity in Turbulent Flows

Up to now we have assumed laminar flow. In that type of flow the
microscopic particles of the medium move along paths that never intersect
one another. An exchange of mass between neighboring paths in gases
takes place only within distances that are of the order of magnitude of
the free-mean path of molecular motion. As we saw before, the "molecular viscosity" of gases is a consequence of that molecular motion.

In *turbulent* flow, a permanent irregular mixing of the paths takes
place. Microscopic particles (turbulent clots) of the flowing medium
undergo irregular side motions in all directions in addition to the main
motion. One is confronted with a kind of magnified picture of the molecular motion of gases. It is also possible, in this case, to speak of a "free-
mean path l_t" of the turbulent motion, which Prandtl calls "mixing
path": it is the length over which individual turbulent clots move as more
or less uniform configurations until they lose their individuality by a
collision and mixing with other clots. Some of these turbulent clots move
perpendicular to the main flow (x direction) with a velocity whose deviation from the time average shall be denoted by $\tilde{v}$. If the time average of
the velocity component u in x direction has a gradient in y direction
($\partial u/\partial y \neq 0$), then the turbulent exchange motion of velocity $\tilde{v}$ results in
a momentum transport in positive or negative y direction. Such an ex-
change of momentum results, as in the case of molecular motion in
laminar flow, in shear stresses in positive or negative x direction between
neighboring fluid domains (cf. Fig. 1.1). These turbulent shear stresses
necessarily obey a relation which corresponds to eq. (1.5):

$$\tau_s = \rho \tilde{v} \, l_t \frac{\partial u}{\partial y}. \tag{1.12}$$

A more detailed investigation of this turbulent exchange process shows
however, as Prandtl [95] was first to recognize, a decisive difference to
the molecular exchange process in laminar flow. The very high velocity $\tilde{v}$
of the molecular motion depends only on the material and the temperature of the medium but not on the velocity u of the main motion (as long
as $\tilde{v} \gg u$). Contrary to this, the relatively small turbulent exchange
velocity $\tilde{v} < u$ depends very strongly on the main motion, for reasons of

continuity. If two turbulent clots move toward one another in y direction with a velocity $\tilde{v}$, then (assuming a 2-dimensional configuration that suffices to estimate the order of magnitudes) two equally large turbulent clots will have to move in x direction with a velocity $\tilde{u}$, also for reasons of continuity. Through the definition of the "mixing path" l_t (Fig. 1.2),

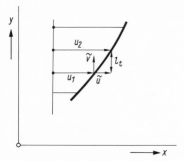

Fig. 1.2.   Illustration of Prandtl's mixing path l_t.

this velocity $\tilde{u}$ is given as the deviation from the time average of the velocity u by

$$\tilde{u} = \pm l_t \frac{\partial u}{\partial y}. \tag{1.13}$$

For reasons of continuity, we also have:

$$\tilde{v} \approx \pm l_t \frac{\partial u}{\partial y} = \tilde{u}. \tag{1.14}$$

The time average of the velocity deviations $\tilde{u}$ and $\tilde{v}$ is then equal to 0 according to their definitions:

$$\bar{\tilde{u}} = \frac{1}{\Delta t} \int_t^{t+\Delta t} \tilde{u} \, dt = 0; \tag{1.15}$$

$$\bar{\tilde{v}} = \frac{1}{\Delta t} \int_t^{t+\Delta t} \tilde{v} \, dt = 0. \tag{1.16}$$

However, the time average $\overline{\tilde{u}\tilde{v}}$ of the product $\tilde{u}\tilde{v}$ at a given location is usually not equal to 0. A measure for the deviation of the time average of the product of the velocities $\tilde{u}$ and $\tilde{v}$ from the product of the time averages of these velocities is given by the "correlation coefficient"

$$K(y) = \frac{\overline{\tilde{u}\tilde{v}}}{\sqrt{\tilde{u}^2} \sqrt{\tilde{v}^2}}. \tag{1.17}$$

K has generally values between 0 and 0.45. For turbulent flow this leads to an important relation for the "apparent" shear stress τ_s in x direction

$$\tau_s = -\rho\overline{\tilde{u}\tilde{v}} = \rho l_t^2 \left|\frac{\partial u}{\partial y}\right|\frac{\partial u}{\partial y} \cdot^* \tag{1.18}$$

The mixing path l_t in eq. (1.18) is defined as a time average, and so is the correlation coefficient. In the vicinity of solid walls we find experimentally that

$$l_t \approx 0.4y. \tag{1.19}$$

It is also found that the apparent shear stress in x direction depends on the square of the velocity gradient. In the laminar case, however, the dependence is linear according to Newton's law.

As the distance from the wall y tends to zero, the fluctuation velocities $\tilde{u}$ and $\tilde{v}$ also go to zero. In every turbulent flow, which is close to the wall, there exists therefore a "laminar sublayer" in which only the molecular velocity is practically significant. This laminar sublayer is generally very thin.

For the shear stress τ in arbitrary (laminar and turbulent) flows, it is therefore possible to write in general

$$\tau_e = (\mu + \mu_s)\frac{\partial u}{\partial y} = \mu_e \frac{\partial u}{\partial y}. \tag{1.20}$$

Here μ is a molecular viscosity, μ_s the "apparent" viscosity, and

$$\mu_e = \mu + \mu_s \tag{1.21}$$

the "effective" viscosity. For μ_s, it follows from eqs. (1.18) and (1.19) that

$$\mu_s = \rho l_t^2 \left|\frac{\partial u}{\partial y}\right| \approx 0.16\rho y^2 \left|\frac{\partial u}{\partial y}\right|. \tag{1.22}$$

It is significant that for large Reynolds numbers, which are necessarily assumed when using boundary layer theory, the apparent shear stresses in fully turbulent flow exceed the Newton stresses generally by orders of magnitude. The "apparent" viscosity μ_s in turbulent flow is therefore usually significantly larger than the "molecular" viscosity μ in laminar flow.

---

*We refrain from writing the square $(\partial u/\partial y)^2$, in order to obtain the correct algebraic sign. Additional details for the problem of turbulence can be found, for example, in Schlichting [111] and the references found there.

1.4 "Apparent" Heat Conductivity in Turbulent Flow

The exchange of mass in turbulent flow leads, besides the exchange of momentum, to an exchange of other state variables, such as for example the enthalpy and thus the temperature T. In turbulent flow the heat exchange is therefore increased, just as though the heat conductivity were larger. Fourier's relation for the heat flow q in y direction, which holds for laminar flow (per unit of time and area)

$$q(x, y) = -\lambda \frac{\partial T}{\partial y}, \tag{1.23}$$

where λ is the molecular heat conductivity, may be written for turbulent flows in analogy to eq. (1.20) in the general form

$$q(x, y) = -(\lambda + \lambda_s) \frac{\partial T}{\partial y} = -\lambda_e \frac{\partial T}{\partial y}, \tag{1.24}$$

where λ_s is the "apparent" and

$$\lambda_e = \lambda + \lambda_s \tag{1.25}$$

is the "effective" heat conductivity. For large Reynolds numbers, [eq. (1.39)], the ratio λ_s/λ is much larger than 1, as is μ_s/μ.

1.5 The Concept of Boundary Layer Thickness

In the motion of a viscous medium along a fixed wall, an important role is played not only by viscosity but also by the attachment of the medium to the wall as a result of the molecular forces of attraction (adhesion). At such a solid wall, the velocity goes down to zero even for arbitrarily small viscosities. From there the velocity increases more or less rapidly to the value to be expected in an ideal fluid for the given contour according to potential theory.

Prandtl [94] recognized that for small viscosities, (for example, in air and water) this domain of increasing velocity extends in practice only to a relatively small distance $y = \delta$ from the wall. Accordingly he called this domain the "boundary layer." It is apparently characteristic for this boundary layer that the shear stresses in eq. (1.1) or (1.20) reach the order of magnitude of the inertial and pressure forces because of the rapid rise of the velocity with y ($\partial u/\partial y$ is very large). The computation of the velocity distribution in this boundary layer from the equilibrium condition for the viscous, inertial, and pressure forces is the subject of "boundary layer theory" (in certain cases buoyant forces also have to be considered).

The velocity inside the boundary layer approaches the free-stream velocity u_δ continuously and in such a way that the higher derivatives remain also continuous. Since, from a theoretical point of view, the boundary layer has to be considered to be of infinite extent in the y direction, it is necessary to conclude that the transition of u into u_δ is asymptotic in nature. In practice, it turns out that the free-stream velocity is reached very rapidly, and the assumption of a finite boundary layer thickness therefore appears justified.

The special case of Hartree profiles shown in Table 3.1 (pp. 77f) illustrates this asymptotic approach very well and shows also why the assumption of a finite boundary layer thickness is justified.

This estimate also shows that the assumption of a finite boundary layer thickness δ is, in principle, not correct. Theoretically viscosity acts to infinite distances y from the wall. In practice, however, the influence of viscosity decreases very rapidly at very small distances from the wall; consequently it appears meaningful to speak of a finite boundary layer thickness δ. In order to define this boundary layer thickness δ uniquely, it has been proposed to correlate it with a velocity u_δ which is equal to 99% of the unperturbed external velocity $u_{\delta p}$, obtained from potential theory:

$$u(x, \delta) = u_\delta(x) = 0.99 u_{\delta p}(x). \tag{1.26}$$

Here $u_{\delta p}$ is the velocity that occurs in pure potential flow at a distance δ from the surface of the contour. For the flow along a flat plate, $u_{\delta p}$ is identical with the unperturbed potential velocity u_∞ at a distance $y = \infty$. In flow contours with displacement action or circulation, $u_{\delta p}$ may differ strongly from $u_\infty (u_{\delta p} > u_\infty)$. Within the framework of a boundary layer theory (except for the case of a flat plate) it is suggested not to choose as the external boundary condition $y \to \infty$, $u \to u_\infty$.

When developing the approximation theory we shall find that the definition of the boundary layer thickness according to eq. (1.26) is unnecessary if one chooses as a measure of the extent of the boundary layer the integrals over the velocity loss or momentum loss within the boundary layer, the so-called *displacement thickness* or *momentum-loss thickness*, respectively. In our future considerations, the boundary layer thickness δ will only play the role of an auxiliary quantity which will, however, prove extremely useful, as for example in the estimates which follow.

1.6 Estimate for the Boundary Layer Thickness

Following the line of thought of Prandtl [94], we shall first estimate how the boundary layer thickness δ develops in plane laminar flow along a solid wall and what order of magnitude this boundary layer reaches. For the sake of simplicity, we consider the flow along a flat plate, that is, a flow without a pressure gradient in the direction of flow (x direction) (Fig. 1.3). In this case only inertial and molecular viscosity forces

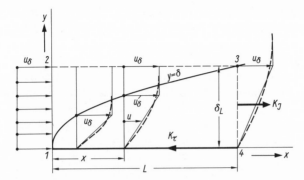

Fig. 1.3. Estimating the development and thickness of a laminar boundary layer in parallel flow along a flat plate (flow without pressure gradient).

are acting within the boundary layer. (Since the pressure forces in a flow with pressure gradient are of the same order of magnitude as the inertial forces, the subsequent estimate will also hold for the more general case of a flow with a pressure gradient in x direction.) The following basic statement can be made about the development of the boundary layer thickness in the direction of flow:

At the leading edge of the plate ($x = 0$), the boundary layer thickness is obviously 0. The masses that were delayed at the beginning of the plate are carried downstream along the flat plate and add to the material already held up there. *The thickness δ of the boundary layer, therefore, increases continuously in the direction of flow.*

We get a better idea about the law that governs the development of the boundary layer thickness δ in a laminar flow by determining roughly the balance of forces in the flow pattern of Fig. 1.3. The boundary layer thickness δ is assumed to rise from the value 0 at the leading edge ($x = 0$) of the flat plate according to the — as yet unknown — law $\delta(x)$ to the value $\delta = \delta_L$ at the end of the plate $x = L$. We do not yet know any details about the velocity u within the boundary layer (the velocity profile). The following boundary conditions are given:

$y = 0$: $u = 0$.

$$y \geq \delta: \quad u = u_\delta; \quad \left(\frac{\partial u}{\partial y}\right)_\delta = 0; \quad \tau_\delta = \left(\mu \frac{\partial u}{\partial y}\right)_\delta = 0. \qquad (1.27)$$

For our estimate it will suffice to assume the velocity distribution $u(x, y)$ between $y = 0$ and $y = \delta$ as being linear in y, that is:

$$u(x, y) = \frac{u_\delta}{\delta(x)} y; \qquad 0 < y < \delta. \tag{1.28}$$

The "asymptotic" value u_δ is reached exactly at the distance $y = \delta$ as indicated in Fig. 1.3.

We now choose a rectangular control section 1, 2, 3, 4, with the plate length L and height δ_L and determine the *viscous force* and the *inertial force* that act upon this rectangle in x direction, the latter as a result of the change in momentum. The condition for the equilibrium of forces will then result in the sought-after boundary layer thickness δ_L and the expression for $\delta(x)$. Since the forces are determined over the entire length L (or x), we must expect the result to be only an estimate.

To find the inertial force that acts on the control section in x direction, we determine the difference between the entering momentum on the left and the leaving momentum on the right, each per unit of width and unit of time. The density ρ and the (molecular) viscosity μ are assumed to be constant in this case. Entering momenta are counted as positive, leaving momenta, as negative.

Across the boundary 1, 2 the following momentum enters in x direction:

$$\rho u_\delta \, \delta_L \cdot u_\delta = \rho u_\delta^2 \delta_L. \tag{1.29}$$

Using eq. (1.28), the following momentum leaves across the boundary 3, 4:

$$-\rho \left(\frac{u_\delta}{\delta_L} \right)^2 \int_0^{\delta_L} y^2 \, dy = -\tfrac{1}{3} \rho u_\delta^2 \delta_L. \tag{1.30}$$

As a result of the displacement action of the boundary layer, a reduced mass flow (= mass per unit of time and per unit of width) leaves across the boundary 3, 4

$$-\int_0^{\delta_L} \rho u \, dy = -\rho \frac{u_\delta}{\delta_L} \int_0^{\delta_L} y \, dy = -\tfrac{1}{2}\rho u_\delta \delta_L. \tag{1.31}$$

With reference to eq. (1.28), this turns out to be exactly one half of the mass flow that enters across the boundary 1, 2. For reasons of continuity, the other half has to leave across the boundary 2, 3, which is parallel to the x axis and in which the velocity u_δ prevails. This mass flow carries the x momentum

$$-\tfrac{1}{2} \rho u_\delta \delta_L \cdot u_\delta = -\tfrac{1}{2}\rho u_\delta^2 \delta_L \tag{1.32}$$

out of the control section.

The average change of momentum in the control section and thus the inertial forces K_J per unit width (in positive x direction) are then given by

$$K_J = \rho u_\delta^\delta \delta_L - \tfrac{1}{3} \rho u_\delta^2 \delta_L - \tfrac{1}{2} \rho u_\delta^2 \delta_L = \tfrac{1}{6} \rho u_\delta^2 \delta_L. \qquad (1.33)$$

It is now necessary to determine the viscous force K_τ per unit width acting on the control section 1, 2, 3, 4 (as a delaying force in negative x direction). This force is the result of adding the local stresses $\tau_w(x)$ along the plate (line $1-4$ in Fig. 1.3). Along line $2-3$ there are no shear stresses, since $(\partial u / \partial y)_\delta = 0$. By Fig. 1.3 and eq. (1.28), the local shear stress (per unit area) $\tau_w(x)$ is given by

$$\tau_w(x) = -\mu \left(\frac{\partial u}{\partial y} \right)_{y=0} = -\mu \frac{u_\delta}{\delta(x)}. \qquad (1.34)$$

In order to determine the unknown average $\bar{\tau}_w$, we introduce into eq. (1.34) an average $\bar{\delta}$ for the boundary layer thickness $\delta(x)$ on the interval $0 < x < L$:

$$\bar{\delta} = c \delta_L, \qquad (1.35)$$

where the numerical factor c assumes values between 0 and 1, thus, on the average, for example, 0.5. The average pulling force along the length of the plate L (per unit width) is then given by

$$K_\tau = -\mu \frac{u_\delta}{c \delta_L} L. \qquad (1.36)$$

The sum of the forces K_J and K_τ has to be 0. From $K_J + K_\tau = 0$, it follows then

$$\tfrac{1}{6} \rho u_\delta^2 \delta_L = \mu \frac{u_\delta}{c \delta_L} L \qquad (1.37)$$

or

$$\frac{\delta_L}{L} = \frac{\sqrt{\dfrac{6}{c}}}{\sqrt{\dfrac{\rho u_\delta L}{\mu}}} \approx \frac{3.5}{\sqrt{R_L}} \quad \text{for} \quad c = 0.5,^* \qquad (1.38)$$

---

*A more precise computation, based on eq. (6.15) of the approximation theory, as outlined in Section 6.2, results in the numerical value 5.64 instead of 3.5. We can see from Fig. 1.3 that the boundary layer thickness δ (defined as the distance from the wall where $u = u_\delta$) based on the actual velocity profile (dashed line) should be slightly larger than in this case, where u is approximated by a straight line.

where

$$R_L = \frac{\rho u_\delta L}{\mu} \tag{1.39}$$

(= Reynolds number, based on the length L of the plate). After a distance x we find for the boundary layer thickness $\delta(x)$ the expression

$$\frac{\delta(x)}{L} = \frac{x}{L}\frac{\delta(x)}{x} \approx \frac{x}{L}\frac{3.5}{\sqrt{\dfrac{\rho u_\delta x}{\mu}}} = 3.5\frac{\sqrt{x/L}}{\sqrt{R_L}}. \tag{1.40}$$

We recognize from eqs. (1.38) and (1.40) that the boundary layer thickness increases in proportion to the root of the dimensionless length x/L and is inversely proportional to the square root of the Reynolds number R_L at every point x/L (hence, also at the end of the plate, $x/L = 1$). A numerical example shall illustrate the order of magnitude of the boundary layer thickness δ_L at the end of the plate. We choose air as the flowing medium with a density

$$\rho = 1.2265 \text{ N sec}^2/\text{m}^4 \qquad (\text{N} = \text{newton})$$

and the viscosity

$$\mu = 1.766 \times 10^{-5} \text{ N sec/m}^2$$

at atmospheric pressure and 20° C, moreover

$$L = 1 \text{ m}$$
$$u_\delta = 15 \text{ m/sec.}$$

Then we have

$$R_L = 10^6 ; \quad \frac{1}{\sqrt{R_L}} = 10^{-3},$$

or

$$\delta_L/L = 3.5 \times 10^{-3}$$
$$\delta_L = 0.0035 \text{ m} = 3.5 \text{ mm.}$$

Thus, the boundary layer thickness δ_L in this example is very small compared with the length L of the plate.

We assume the same length for the plate and choose water as the flowing medium with the values $\rho = 996 \text{ N sec}^2/\text{m}^4$, $\mu = 100 \cdot 10^{-5} \text{ N sec/m}^2$.

The same boundary layer thickness δ_L is already reached at a velocity of 1 m/sec.

Even when the velocities or plate lengths are smaller by two orders of magnitude than in this example ($R_L = 10^4$ instead of 10^6), the boundary layer thickness at the end of the plate is always only a few percent of the total length of the plate L.

This means that in fluid mechanics with air (or with gases in general) or water as the flowing media, the Reynolds numbers usually are so large that the boundary layer thickness is small compared to the length x of the flow. It can be shown that this result as regards the connection between Reynolds number and boundary layer thickness also applies to turbulent boundary layers, at least within the same order of magnitude [cf., for example, Section 6.2.3, eq. (6.22)], but it applies also to flows with pressure gradients in direction of flow. This fact allows for important conclusions concerning the theoretical treatment of flows with large Reynolds numbers which will be formulated more precisely in the following Section.

1.7 Conclusions Derived from the Small Boundary Layer Thickness (Prandtl's Boundary Layer Simplifications)

If we neglect viscosity in a parallel flow along a plane wall (stratified flow), there exists in the immediate vicinity of the wall only a velocity component u parallel to the wall (x component). In actual flow under the influence of viscosity and the no-slip condition at the wall, the streamlines which before were parallel to the wall are now somewhat displaced away from the wall for reasons of continuity. By this displacement the flowing medium acquires a velocity component v in y direction. If the wall is impermeable, both velocity components v and u are equal to zero at the wall ($y = 0; u = v = 0$). At the outer edge of the boundary layer, that is, at $y = \delta$, the velocity v (as long as $\partial u/\partial y = 0$ for $y = \delta$)* reaches its maximum value v_δ. Based on the estimates in the previous section, we shall compute now this value v_δ.

First we note that the result of this estimate for the maximum "displacement velocity" v_δ (just as for the boundary layer thickness δ) applies also to flows with pressure gradients, at least within an order of magnitude.

*Actually v increases outside $y = \delta$ still further, since $\partial u/\partial y$ is not exactly 0 there. For the subsequent estimate, we may, however, assume a finite boundary layer thickness δ with $\partial u/\partial y = 0$ for $y > \delta$.

According to Fig. 1.3, the displaced mass flow m_v (for constant density $\rho = \rho_\delta = $ const) at a distance x from the leading edge is given by

$$m_v = \rho_\delta u_\delta \int_0^\delta \left(1 - \frac{u}{u_\delta}\right) dy = \rho_\delta u_\delta (\delta_1)_u.^* \tag{1.41}$$

It is customary to refer the length $(\delta_1)_u$ as "displacement thickness." $(\delta_1)_u$ is the distance from the wall by which the potential flow of constant velocity u_δ (for constant density $\rho = \rho_\delta$) appears to be displaced from the wall. Assuming a linear relation for $u(y)$ inside the boundary layer $[0 < y < \delta]$ as given by eq. (1.28) we obtain

$$(\delta_1)_u = \int_0^\delta \left(1 - \frac{y}{\delta}\right) dy = \delta - \frac{1}{2} \frac{\delta^2}{\delta} = \frac{1}{2} \delta . \tag{1.42}$$

For a linear velocity profile, we thus obtain

$$\frac{(\delta_1)_u}{\delta} = \tfrac{1}{2} . \tag{1.43}$$

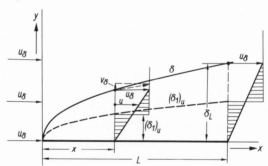

Fig. 1.4. Estimating the order of magnitude of the displacement velocity v_δ at the edge $y = \delta$ of the boundary layer.

In Fig. 1.4 we have, in addition to the graph of $\delta(x)$ from Fig. 1.3, plotted the graph of the displacement thickness $(\delta_1)_u(x)$. The displacement velocity v_δ sought at the point $y = \delta$ is then obviously given by

$$v_\delta = u_\delta \frac{d(\delta_1)_u}{dx} \tag{1.44}$$

---

*Anticipating a more general definition for δ_1, to be used later for the displacement thickness for variable densities, we add in this definition the subscript u to indicate that the displacement thickness in this special case depends only on the velocity profile u/u_δ and not as yet on the density profile ρ/ρ_δ.

or, because of the proportionality between the total boundary layer thickness δ and $(\delta_1)_u$, see eq. (1.43),

$$v_\delta = \tfrac{1}{2} u_\delta \frac{d\delta}{dx}. \tag{1.45}$$

Based on Fig. 1.4, we may estimate the gradient $d\delta/dx$, if we exclude the small domain $0 < x < x_0$ [for $x \to 0$, we have from eq. (1.40) $d\delta/dx \to \infty$], as follows:

$$\frac{d\delta}{dx} \approx \frac{\delta_L}{L}. \tag{1.46}$$

Thus we arrive at the following estimate for the ratio of the displacement velocity v_δ at the edge of the boundary layer to the velocity component u_δ parallel to the wall, under consideration of eq. (1.38):

$$\frac{v_\delta}{u_\delta} \approx \frac{1}{2} \frac{\delta_L}{L} \approx \frac{1.75}{\sqrt{R_L}} \ll 1 \quad \text{for} \quad R_L \gg 1. \tag{1.47}$$

For large Reynolds numbers, $R_L \gg 1$, the displacement velocity v_δ in y direction (perpendicular to the fixed wall) which is caused by the boundary layer is very small compared with the x component u_δ of the velocity at the point $y = \delta$. Since $u = v = 0$ at $y = 0$, we have, moreover, within the entire boundary layer ($0 < y < \delta$) for $R_L \gg 1$

$$0 < v < v_\delta \tag{1.48}$$

and thus also

$$\frac{v}{u_\delta} \ll 1. \tag{1.49}$$

For flows along a plane wall, this result allows the following important conclusion about the character of the *static pressure* $p(x, y)$ within the boundary layer (except for the narrow domain $0 < x < x_0$ in the immediate vicinity of the leading edge $x = 0$): the motion of the medium with a displacement velocity v_δ perpendicular to the wall at a fixed point x, may be considered to have been caused by a pressure difference

$$\Delta p = p(x, 0) - p(x, \delta) = p_w - p_\delta \tag{1.50}$$

between the wall ($y = 0, p = p_w$) and the edge ($y = \delta, p = p_\delta$) of the boundary layer.

Since this motion in y direction, owing to the smallness of v, has only very small velocity gradients $\partial v/\partial x$ and thus very small viscous forces $\mu(\partial v/\partial x)$ in the y direction, we may assume that for this motion only the inertial and pressure forces are practically significant. We may therefore write a Bernoulli equation for this motion (provided it is considered independent of the main flow in x direction for purposes of this estimate) as follows:

$$p_\delta + \frac{\rho}{2}v_\delta^2 = p_w + \frac{\rho}{2}v_w^2 = p_w \tag{1.51}$$

(because $v_w = 0$). Therefore, we have

$$p_w - p_\delta = \Delta p = \frac{\rho}{2}v_\delta^2. \tag{1.52}$$

The pressure difference Δp within the boundary layer divided by twice the stagnation pressure ρu_δ^2 of the external flow, after using eq. (1.47), is then

$$\frac{\Delta p}{\rho_\delta u_\delta^2} = \frac{v_\delta^2}{u_\delta^2} \approx \frac{3}{R_L}. \tag{1.53}$$

The pressure difference Δp within the boundary layer may also be considered to have been caused by the different curvature of the streamlines at the wall (curvature radius $\mathfrak{R}$ of the flat plate $= \infty$) and the outer edge of the boundary layer. The radius of curvature of the streamline $y = \delta$ is given by

$$\mathfrak{R} = \frac{\left[1 + \left(\dfrac{d\delta}{dx}\right)^2\right]^{3/2}}{d^2\delta/dx^2} \approx \frac{1}{d^2\delta/dx^2}. \tag{1.54}$$

For the centrifugal force K_Z (per unit area), which a particle whose mass is proportional to $\rho u_\delta \delta$ experiences during a motion along the streamline, we find

$$K_Z \sim \frac{\rho u_\delta^2 \delta}{\mathfrak{R}} = \rho u_\delta^2 \frac{d^2\delta}{dx^2}\delta, \tag{1.55}$$

that is

$$\frac{K_Z}{\rho u_\delta^2} \sim \delta \frac{d^2\delta}{dx^2}. \tag{1.56}$$

Along the streamline $y = 0$, we have $\mathfrak{R} = \infty$ and, thus, $K_z = 0$. Since equilibrium prevails, there is a relative pressure difference between $y = 0$ and $y = \delta$ which is given by $\Delta p/\rho u_\delta^2 = -K_Z/\rho u_\delta^2$.

Using the estimate (1.46) for $d\delta/dx$ with $x \sim L$, and eq. (1.38), we find

$$-\frac{\Delta p}{\rho u_\delta^2} \sim \frac{K_Z}{\rho u_\delta^2} \sim \left(\frac{\delta_L}{L}\right)^2 \sim \frac{1}{R_L}. \tag{1.57}$$

The algebraic sign of Δp in definition (1.52) is chosen such that $p_\delta > p_w$, that is, the reverse as in eq. (1.53). The order of magnitude of $\Delta p/\rho u_\delta^2$ is the same in both cases, however.

Without doubt both forces, as given by eqs. (1.53) and (1.57), are present simultaneously. Which of the two is more significant has to be decided in each individual case. For present purposes it suffices to know the order of magnitude, which is unaffected when adding and subtracting the two forces given in eqs. (1.53) and (1.57).

In any case, for very large Reynolds numbers (R_L larger than about 10^3) which occur preponderantly in fluid mechanics, the relative pressure difference $\Delta p/\rho u_\delta^2$ in y direction is very small. The flow within the boundary layer is therefore practically determined by the equation of motion in x direction and by the continuity condition (from which follows the displacement effect). Again, the very small domain in the vicinity of the leading edge $x = 0$ has to be excluded.

The equation of motion in y direction for a plane wall or for very large radii of curvature $\Re$ at the wall ($x/\Re \ll 1$) within the boundary layer can, therefore, be reduced to the statement:

$$\frac{\partial p}{\partial y} = 0 \quad \text{for} \quad 0 < y < \delta. \tag{1.58}$$

From this follows that the pressure p_w at the wall ($y = 0$) equals the pressure p_δ at the edge ($y = \delta$) of the boundary layer. The static pressure p_δ of the potential flow and its gradient dp_δ/dx are therefore imposed upon the entire boundary layer.

If the boundary layer is thin, that is, its displacement effect is small, then the static pressure $p_w(x) = p_\delta(x) = p(x)$ along the surface of the given body, for example the surface of an airplane wing, may be computed without considering the boundary layer phenomena, strictly from classical hydrodynamics of an ideal fluid using well-known methods (conformal mapping for two-dimensional flow, singularity methods for two- and three-dimensional flow). If necessary, the displacement effect can be accounted for by iteratively correcting the effective contour (the actual contour plus the displacement thickness) (cf. Section 5.4.2).

On the basis of these results for our estimates for viscous flow with very large Reynolds numbers it is permissible to split each flow problem into two separate problems which are much simpler to solve separately:

(a) the problem to determine the *pressure function p(x)* for a given profile according to well-known methods of classical hydrodynamics of ideal fluids:

(b) a pure *boundary layer problem* for which only the equation of motion for the x component of the forces parallel to the wall is significant, and the pressure forces in that direction of motion (the pressure gradient dp/dx) can be obtained from the potential theoretical computation.

The velocity $u_\delta(x)$ at the edge of the boundary layer and its derivative with respect to x can be obtained from the Bernoulli equation

$$\int \frac{dp}{\rho_\delta} + \frac{u_\delta^2}{2} = \text{const.} \tag{1.59}$$

$$\frac{dp}{dx} = -\rho_\delta u_\delta \frac{du_\delta}{dx}. \tag{1.60}$$

The relations (1.59) and (1.60) are also valid for compressible isentropic flow (cf. Section 2.3.2). One should note that in these general relations the density ρ_δ at the edge of the boundary layer has to be assumed a function of δ, $\rho_\delta(x) = \rho(x, \delta)$.

At the leading edge $x = 0$ of a flat plate (and more generally of bodies with vanishing angle at the leading edge) Prandtl's boundary layer simplifications, as described, no longer apply. The singularity at the point $x = 0$ does not invalidate the computations in practical cases, certainly not in computations based on approximation theories. For theoretical details about this singularity at $x = 0$, we refer to the literature (for example, Schlichting [111]).

2 Short Summary of the Exact Theory of Boundary Layers for Flow and Temperature

2.1 Equation of Motion for the Boundary Layer of Flow (*Direct Derivation of Prandtl's Boundary Layer Equation*)

Based on the results of Sections 1.6 and 1.7, which are referred to as Prandtl's boundary layer simplifications, we are now able to formulate the equation of motion for the boundary layer of flow directly (without referring to the Navier-Stokes equations). Here we are limiting ourselves to two-dimensional stationary flows along a fixed wall without sources and sinks. This wall may be curved. However, the local radius of curvature $\Re$ is assumed to be so large compared to the boundary layer thickness δ that the pressure gradients in perpendicular direction to the flow, which are caused by this curvature (centrifugal forces), are negligible. We also assume that the buoyant forces are negligible against all other forces. These assumptions are generally satisfied with good accuracy (exception: for example, flows with a free surface). The x direction of our Cartesian coordinate system is assumed to coincide with the tangent at the wall along which the viscous medium flows (Fig. 2.1).

We consider now an infinitesimal rectangular control section with sides dx, dy and the width 1 within the boundary layer and assume that for a

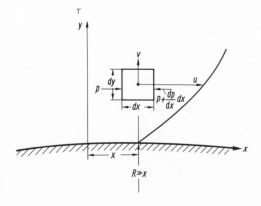

Fig. 2.1. Choice of a coordinate system for two-dimensional stationary flow along an almost flat wall.

large Reynolds number the Prandtl simplifications are satisfied here. Within the boundary layer, we denote by u the velocity in x direction and by v the velocity in y direction. It is assumed that v is much smaller than u.

We first formulate the condition of *continuity* for the flow. The following mass flow enters the control section (Fig. 2.1) from the left per unit of time (and per unit width)

$$\rho u \, \mathrm{d}y, \tag{2.1}$$

and on the right-hand side the mass flow

$$\left(\rho u + \frac{\partial(\rho u)}{\partial x} \, \mathrm{d}x \right) \mathrm{d}y \tag{2.2}$$

leaves. The mass flow therefore changes within the control section by the amount

$$\frac{\partial(\rho u)}{\partial x} \, \mathrm{d}x \, \mathrm{d}y. \tag{2.3}$$

The control section is assumed to be free of sources and sinks, and thus for reasons of continuity this change in the mass flow has to be accompanied by a mass flow in y direction through the sides $\mathrm{d}x$ of the control section. In y direction, however, the mass flow per unit of time from below is given by

$$\rho v \, \mathrm{d}x. \tag{2.4}$$

At the upper edge, the mass flow

$$\left(\rho v + \frac{\partial(\rho v)}{\partial y} \, \mathrm{d}y \right) \mathrm{d}x \tag{2.5}$$

leaves the control section. The change in the flow of mass in y direction is thus given by

$$\frac{\partial(\rho v)}{\partial y} \, \mathrm{d}y \, \mathrm{d}x. \tag{2.6}$$

The sum of the changes (2.3) and (2.6) has to be 0:

$$\frac{\partial(\rho u)}{\partial x} + \frac{\partial(\rho v)}{\partial y} = 0. \tag{2.7}$$

We now formulate the condition for the *equilibrium of* those *forces* that act on the control section in x direction. (The forces acting on the control section in y direction are negligible within the framework of Prandtl's boundary layer simplifications, which are assumed valid.)

In principle, we have to consider the following forces: *inertial forces, pressure forces, viscous forces* (shear stresses), and *buoyant forces*. In the following, we limit ourselves to flows without buoyant forces.

In general, within the domain of validity of Prandtl's boundary layer simplifications, the buoyant forces caused by density differences within the boundary layer are negligible compared with the other forces. This also holds for flows of compressible media with large density differences in the boundary layer.

The *inertial force* acting on the control section is found as the difference between the entering and leaving momenta per unit of time (and per unit width). A momentum entering in x direction shall be denoted by a positive sign.

The following momentum enters the control section from the left per unit of time and width

$$\rho u \cdot u \, dy = \rho u^2 \, dy. \tag{2.8}$$

On the right hand side, the momentum

$$-\left(\rho u^2 + \frac{\partial(\rho u^2)}{\partial x} \, dx\right) dy \tag{2.9}$$

leaves. The change of momentum that results from the change in mass flow as given by eq. (2.3) is thus

$$-\frac{\partial(\rho u^2)}{\partial x} \, dx \, dy \tag{2.10}$$

and corresponds to a force (per unit width) which acts in the negative x direction.

The expression (2.10), however, does not give the entire inertial force acting in x direction. As a result of the continuity equation, some of the mass flow has been diverted into the y direction (or, is drawn in from there), and thus an additional x momentum results. At the lower edge of the control section, the mass flow $\rho v \, dx$ has the following x momentum

$$\rho v u \, dx. \tag{2.11}$$

The same mass flow, when leaving the upper edge, carries the x momentum

$$-\left(\rho v u + \frac{\partial(\rho v u)}{\partial y} \, dy\right) dx. \tag{2.12}$$

Thus, because of the mass flow through the control section in y direction an additional change in the x momentum, of magnitude

$$- \frac{\partial(\rho v u)}{\partial y} \, dy \, dx, \tag{2.13}$$

results.

For the total change of the x momentum (that is, the inertial force in x direction) we find, after adding the expressions (2.10) and (2.13),

$$- \left(\frac{\partial(\rho u^2)}{\partial x} + \frac{\partial(\rho v u)}{\partial y} \right) dx \, dy. \tag{2.14}$$

After differentiation in the expression (2.14), we obtain

$$- \left[\rho u \frac{\partial u}{\partial x} + \rho v \frac{\partial u}{\partial y} + u \frac{\partial(\rho u)}{\partial x} + u \frac{\partial(\rho v)}{\partial y} \right] dx \, dy. \tag{2.15}$$

The last two terms in expression (2.15) are exactly the equation of continuity (2.7) multiplied by $u \, dx \, dy$, which vanishes identically.

The *inertial force* in x direction can now be written in the final form

$$- \left(\rho u \frac{\partial u}{\partial x} + \rho v \frac{\partial u}{\partial y} \right) dx \, dy. \tag{2.16}$$

The *pressure force* acting on the control section in x direction follows simply from the externally induced pressure, given by potential theory, as the difference

$$p \, dy - \left(p + \frac{dp}{dx} dx \right) dy = - \frac{dp}{dx} dx \, dy \tag{2.17}$$

(cf. Fig. 2.1). The pressure force is positive when $dp/dx < 0$ (that is, for decreasing pressure in the direction of flow).

Finally, we have to formulate the *viscous force*, which acts on the control section in x direction. At the lower edge of the control section (length dx) the shear stress $\tau = \mu \, \partial u / \partial y$ [eq. (1.1)] results in the retarding (negative) force per unit width

$$- \tau \, dx \tag{2.18}$$

At the upper edge, the accelerating viscous force (cf. Sect. 1.2 and Fig. 1.1) is found as

$$\left(\tau + \frac{\partial \tau}{\partial y} dy \right) dx.$$

The viscous force, which acts on the control section (an accelerating force when $\partial \tau / \partial y$ is positive) is thus

$$\frac{\partial \tau}{\partial y} \, dy \, dx. \tag{2.19}$$

The sum of all forces (per unit of width) given by eqs. (2.16), (2.17), and (2.19) must be 0. This, after dividing by $dx \, dy$, leads to Prandtl's boundary layer equation:

$$\rho u \frac{\partial u}{\partial x} + \rho v \frac{\partial u}{\partial y} = -\frac{dp}{dx} + \frac{\partial \tau}{\partial y}, \tag{2.20}$$

where τ is defined by eq. (1.20), and the continuity equation

$$\frac{\partial (\rho u)}{\partial x} + \frac{\partial (\rho v)}{\partial y} = 0 \tag{2.21}$$

holds.

It should be noted here that the simultaneous system of eqs. (2.20), (2.21) was derived without restricting assumptions about the density ρ and the viscosity μ which appears in the expression for the shear stress τ for laminar boundary layers according to eq. (1.1). This system of equations is therefore generally valid in flows with variable material properties — for example, for the boundary layer flow in a compressible medium.

The system of eqs. (2.20), (2.21) is also valid for turbulent boundary layers if the variables u, v and ρ are interpreted as time averages and the effective viscosity μ_e is taken as the sum of the molecular viscosity μ and the apparent viscosity μ_s which, for example, may be empirically given by eq. (1.22). For reasons of simplicity, we shall denote the shear stress τ_e by τ throughout the remainder of this book.

Strictly speaking, in this case, eq. (2.20) has to be augmented on the left hand side by the term

$$\frac{\partial}{\partial x} [\rho (\overline{\tilde{u}^2} - \overline{\tilde{v}^2})].$$

Except in special cases (when approaching a point of flow separation), this term, however, is negligible [see for example eq. (1.14)].

A physically meaningful boundary condition for eqs. (2.20), (2.21) is in every case (for an impermeable wall)

$$y = 0: \quad u = v = 0 \tag{2.22}$$

(no-slip condition for the flow at the wall). An additional boundary condition has to express the fact that for a sufficiently large distance y from the fixed wall, the boundary layer flow has to approach the "external flow" which, according to the estimates in Section 1.6, is practically free of any influence from viscosity. In theory, the influence of viscosity does not vanish until the distance from the wall $y = \infty$ is reached (the boundary layer thickness δ is then actually infinite) so that, strictly speaking, this "outer" boundary condition has to be written as

$$y \to \infty: \quad u \to u_\infty. \tag{2.23}$$

As discussed already in Section 1.5, the velocity in the free-stream flow may still change with the distance from the wall for purely potential theoretical reasons — in particular, in flow problems with strong pressure gradients in x direction. From a physical point of view it is thus more reasonable to require that the velocity $u = u_{\delta p}$ be reached at a finite distance from the wall, which is of the order of magnitude of the boundary layer thickness, as defined in Section 1.5. This velocity would be present in pure potential flow at a distance $y = \delta$. It is possible that $u_{\delta p}$ can differ considerably from u_∞.

So long as Prandtl's boundary layer theory is valid, that is, for large Reynolds numbers, this problem of the outer boundary condition is of no practical significance, because the solution of the system of eqs. (2.20), (2.21) shows that the asymptotic transition from u into $u_{\delta p}$ occurs very suddenly at a small distance from the wall $y \approx \delta$, where δ is taken from eq. (1.40). It is therefore possible to work formally with the boundary condition (2.23) by putting $u_\infty = u_{\delta p}$. In the following, we write for the sake of simplicity $u_{\delta p} = u_\delta$.

The mathematical problem as to whether the system of eqs. (2.20), (2.21), together with the boundary conditions (2.22), (2.23), has a solution remains unanswered for the general case. In the case of constant material properties ρ and μ (which reduces to a single nonlinear partial differential equation of second order for the unknowns u and v) the existence and uniqueness of the solution have been proved by Oleinik [84]. (Cf. also Nickel [81].)

We shall discuss here briefly the so-called "similar solutions," since they will play an important role later when we discuss approximate solutions. For specific questions we refer the reader to the more detailed treatment in Schlichting [111] of these and other exact solutions and methods of solving.

2.2 Exact Solutions of Prandtl's Boundary Layer Equation

2.2.1 "Similar Solutions" of Prandtl's boundary layer equation
for constant material properties
in laminar case

In answering the question whether there are any solutions of the system of eqs. (2.20), (2.21), the pressure gradient as given by potential theory, and thus, the pressure function $p(x)$, plays an important role. The free-stream flow is governed by Bernoulli's equation

$$\frac{\rho_\delta}{2} u_\delta^2 + p = \text{const.} \qquad (2.24)$$

or, in compressible flow [cf. Section 2.3.3.2, eq. (2.96)],

$$\frac{u_\delta^2}{2} + \int \frac{\mathrm{d}p}{\rho_\delta} = \text{const.} \qquad (2.25)$$

The pressure function $p(x)$ may be replaced by the velocity distribution $u_\delta(x)$ and the pressure gradient $\mathrm{d}p/\mathrm{d}x$ may be replaced by the velocity gradient $\mathrm{d}u_\delta/\mathrm{d}x$. From eqs. (2.24) and (2.25) follows by differentiation for incompressible and compressible flows:

$$\frac{\mathrm{d}p}{\mathrm{d}x} = -\rho_\delta u_\delta \frac{\mathrm{d}u_\delta}{\mathrm{d}x}. \qquad (2.26)$$

As first shown by Falkner and Skan [25] in 1931 and later by Goldstein [37] and Mangler [68], the system of eqs. (2.20), (2.21) can be solved in the laminar case for external flows $u_\delta(x)$ of the type

$$u_\delta(x) \sim x^m, \qquad (2.27)$$

where m is quite arbitrary, by a "similarity transformation" to a more easily solvable ordinary differential equation.* It is characteristic for the solution of this ordinary differential equation, which was first given by Hartree [43], that for fixed (positive or negative) values of m all velocity profiles $u(x, y)$ which develop in the direction flow are similar to each other, i.e., they differ only by multipliers (depending on x) that affect the distance from the wall y and the velocity u. The exponent m assumes the role of a shape parameter of the velocity profile.

*There are a few other types of potential flows for which similar solutions can be obtained in boundary layer theory. We refer here to the extensive treatment of this topic in Schlichting [111].

2.2.1.1 Special case of similar solutions: boundary layer along a flat plate for constant material properties. For $m = 0$, the case of a flow along a flat plate: $u_\delta(x) = $ const., $du_\delta/dx = 0$; $dp/dx = 0$ is given. For this simple case, Blasius [4], Prandtl's co-worker, in 1907 found the similarity transformation for the system (2.20), (2.21) on the basis of the boundary layer simplifications. The physical considerations which lead to this transformation are based on the estimates about the development of the boundary layer thickness, eq. (1.40). Since these considerations are also typical for the more general similarity transformation (for values of m other than 0), we shall briefly touch on these ideas here.

From eq. (1.40), it follows that the boundary layer thickness δ develops according to the law

$$\delta(x) \sim \frac{x}{\sqrt{R_x}} \; ; \quad R_x = \frac{\rho u_\delta x}{\mu} \tag{2.28}$$

as a function of the distance x. It is therefore suggested to measure the distance from the wall in units of $\delta(x)$, according to eq. (2.28), (or a multiple thereof), i.e., to introduce a dimensionless distance by

$$\eta = \frac{y}{c\,\delta(x)} = \frac{y}{cx}\sqrt{R_x} = \frac{y}{cx}\sqrt{\frac{x}{L}}\sqrt{R_L}. \tag{2.29}$$

A dimensionless velocity, in this case, where $u_\delta(x) = $ const., is given by the ratio $u(x, y)/u_\delta(x)$. When eliminating the velocity v by using eq. (2.21) (for constant density ρ) in the form

$$v = -\int_0^y \frac{\partial u}{\partial x}\,dy = -\frac{\partial}{\partial x}\left(\int_0^y u\,dy\right), \tag{2.30}$$

we find it more suitable to use instead of the dimensionless velocity u/u_δ the dimensionless stream function

$$f(\eta) = \int_0^\eta \frac{u}{u_\delta}\,d\eta. \tag{2.31}$$

Introducing eqs. (2.29) and (2.31) into eq. (2.20), we find after short and elementary computations in the case of constant material properties ρ and μ, and with $c = \sqrt{2}$, the ordinary differential equation of third order in f (where $'$ indicates differentiation with respect to η).

$$f''' + ff'' = 0, \tag{2.32}$$

with the boundary conditions

$$\eta = 0 \ (y = 0): \quad f' = \frac{u}{u_\delta} = 0, \quad f = 0, \tag{2.33}$$

$$\eta \to \infty \ (y \to \infty \quad \text{or} \quad R_L \to \infty): \quad f' = 1. \tag{2.34}$$

The solution for $f(\eta)$ and thus for the velocity profile $f'(\eta) = u/u_\delta$ is therefore independent of the distance x; that is, all velocity profiles in a laminar boundary layer along a flat plate are similar to each other. The solution of eq. (2.32) cannot be given in closed form. The numerical solution, however, does not encounter fundamental difficulties with today's computers. Blasius [4], though, had to use rather involved methods of solution in his days.

Special attention should be given to a proposition by Piercy and Preston [89] because of its general usefulness for other boundary layer problems. Developed in 1936, their proposal suggests an iterative solution of eq. (2.32) by putting

$$f'' = F \tag{2.35}$$

and assuming a 0th approximation $f^{(0)}(\eta)$ as known. Equation (2.32) is then written as

$$\frac{dF}{d\eta} + f^{(0)}F = 0 \tag{2.36}$$

or

$$\frac{dF}{F} = -f^{(0)} \, d\eta. \tag{2.37}$$

After integrating eq. (2.37) once, we obtain

$$F = c_1 \exp\left[-\int_0^\eta f^{(0)}(\eta) \, d\eta \right]; \tag{2.38}$$

after another integration, with $f' = u/u_\delta$,

$$\frac{u(\eta)}{u_\delta} = c_1 \int_0^\eta \exp\left[-\int_0^\eta f^{(0)} \, d\eta \right] d\eta + c_2; \tag{2.39}$$

and after using the boundary conditions (2.33) and (2.34) to determine the integration constants c_1 and c_2:

$$f'(\eta) = \frac{u}{u_\delta} = \frac{\displaystyle\int_0^\eta \exp\left[-\int_0^\eta f^{(0)}(\eta) \, d\eta \right] d\eta}{\displaystyle\int_0^\infty \exp\left[-\int_0^\eta f^{(0)}(\eta) \, d\eta \right] d\eta}. \tag{2.40}$$

The method converges well. Even if ones starts with $f^{(0)} = \eta$, i.e., with $u = u_\delta = $ const., the exact solution for the function $f'(\eta)$ is obtained within a fraction of one percent after a few iteration steps, even in close proximity to the wall. A proof of convergence for this iteration method was given by Weyl [154]. Careful measurements by J. Nikuradse [83], among others, confirm the results of the theory with respect to the shape as well as to the similarity of the velocity profiles at different lengths x.

For the local shear stress $\tau_w = \mu(\partial u/\partial y)_{y=0}$ Blasius found

$$\frac{2\tau_w}{\rho u_\delta^2} = c_{fi} = \frac{0.664}{\sqrt{R_x}} = \frac{0.664}{\sqrt{x/L}} \cdot \frac{1}{\sqrt{R_L}} \left(R_x = \frac{\rho u_\delta x}{\mu} ; \quad R_L = \frac{\rho u_\delta L}{\mu} \right).$$

(2.41)

For the drag coefficient C_{Fi} of the flat plate with length L (both for one sided and double sided exposure to flow) the integration of c_{fi} (from $x = 0$ to $x = L$) yields

$$C_{Fi} = \frac{W}{\frac{\rho}{2} u_\delta^2 F} = \int_0^1 c_{fi}\, d\frac{x}{L} = 0.664 \cdot \frac{1}{\sqrt{R_L}} \int_0^1 \frac{dx/L}{\sqrt{x/L}} = \frac{1.328}{\sqrt{R_L}}$$

(2.42)

(F = area of the plate exposed to flow, W = frictional force acting on F). This numerical value for the total drag coefficient C_{Fi} of the flat plate is in best agreement with the experiment.

2.2.1.2 General case of similar solutions. According to Mangler [68], the system (2.20) and (2.21) can be transformed into an ordinary differential equation for positive or negative exponents m in eq. (2.27) if one introduces the following dimensionless quantities for the distance from the wall and the stream function:

$$\eta = \sqrt{\frac{m+1}{2}} \cdot \frac{y}{x} \sqrt{R_x},$$

(2.43)

$$f(\eta) = \int_0^\eta \frac{u}{u_\delta}\, d\eta.$$

(2.44)

If one replaces m by

$$m = \frac{\beta^*}{2 - \beta^*}$$

(2.45)

this differential equation becomes

$$f''' + ff'' + \beta^*(1 - f'^2) = 0$$

(2.46)

[boundary conditions as in eq. (2.32)].* One recognizes that this equation reduces to (2.32) for $\beta^* = 0$, $m = 0$.

Hartree [43] numerically computed solutions of this equation for the interval $-0.1988 < \beta^* < 2.00$ which corresponds to $-0.0904 < m < \infty$. Table 3.1 (pp. 77–79) shows the velocity profiles $f'(\eta) = u/u_\delta$ for selected positive and negative values $\beta^*(m)$. Positive values of β^* and m, according to eq. (2.27), correspond to accelerated external flow, negative values to retarded flow. The value $\beta^* = -0.1988$ ($m = -0.0904$) corresponds to the velocity profile with a tangent at the wall $f''(0) = 0$, a fact which is of significance in the investigation of flow separation.

In Fig. 2.2 some of the velocity profiles of Table 1 are represented graphically. It is characteristic for all velocity profiles in retarded

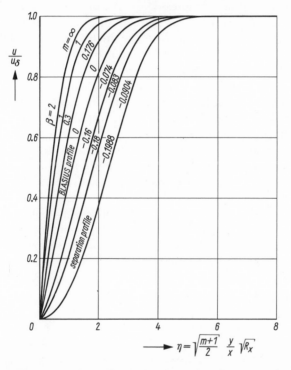

Fig. 2.2. "Similar" solutions of Prandtl's boundary layer equation for laminar boundary layer in incompressible flow. Some velocity profiles of Hartree's numerical solution are shown (see also Table 3.1; read β^* instead of β).

*Kwang-Tzu Yang [61] suggested an iterative solution for this equation (as well as for compressible flow) along the lines of the principle suggested in eqs. (2.35) through (2.40). A generally valid proof of convergence is still missing.

flow (for $\beta^* < 0$, $m < 0$) to have turning points, i.e., the quantity $f''' = \partial^2(u/u_\delta)/\partial\eta^2$ changes its algebraic sign (from positive values in the vicinity of the wall through zero at the turning point to negative values in the outer domain of the velocity profile). For accelerated flow ($\beta^* > 0$, $m > 0$) the velocity profiles are fuller than the Blasius profile for the flat plate, and only negative values of f''' occur (for the Blasius profile we have $f''' = 0$ at the wall.)

The solutions of eq. (2.46) therefore yield a great deal of information about the basic properties of laminar boundary layers in accelerated and retarded flow (for decreasing and increasing pressure). It seems important to point out the following physical facts:

The estimates in Sections 1.6 and 1.7 show that the static pressure $p(x)$ in the free-stream flow is impressed on the boundary layer all the way down to the wall ($y = 0$). The pressure gradient dp/dx at the edge of the boundary layer is therefore also present in the same magnitude along the fixed wall. In the free-stream flow (up to the edge of the boundary layer), there is always enough kinetic energy when a pressure rise occurs to develop the pressure field as given by Bernoulli's law, eq. (2.24), such that in the most unfavorable case for the flow (at a stagnation point) the velocity goes down to zero but can never be forced to reverse its direction. Quite a different situation exists within the boundary layer where the work of the shear stresses consumes kinetic energy and is transformed into heat. Within the boundary layer there is only a limited amount of kinetic energy available to overcome a pressure increase forced on the boundary layer from outside. In the immediate vicinity of the wall, where the kinetic energy goes to zero, it is therefore possible, in principle, that the flowing material may move against the direction of the free-stream main flow under the influence of a pressure increase (a positive pressure gradient, $dp/dx > 0$). But this means flow separation.

We state: flow separation is only possible through the intermediary of a boundary layer in which kinetic energy is reduced against the free-stream flow (potential flow) as the result of the work of shear stresses.

Before flow separation occurs in a domain of increasing pressure, the viscosity counteracts separation (this is at first glance a paradox): the outer parts of the boundary layer exert a retarding force in the direction of flow on the parts closer to the wall, where reverse flow is about to emerge. Under the influence of this retarding force the velocity profile assumes a concave shape in the vicinity of the wall, that is with positive $\partial^2u/\partial y^2$, while in the outer part of the boundary layer (because of the always asymptotic transition of the velocity to the value u_δ of the free-stream flow) the convex shape with negative values of $\partial^2u/\partial y^2$ remains. Between those two domains $\partial^2 u/\partial y^2$ assumes the value 0: velocity profiles in domains of increasing pressure therefore have always a turning point in the domain $0 < y < \delta$, where $\partial^2 u/\partial y^2 = 0$ (cf. Fig. 2.2).

When the pressure decreases in the direction of flow, the negative pressure gradient ($\mathrm{d}p/\mathrm{d}x < 0$) has an accelerating effect on the particles in the vicinity of the wall. The velocity profiles of the boundary layer become fuller in this case than they were without the influence of the negative pressure gradient. In flows of the type $u_\delta \sim x^m$ only velocity profiles with negative $\partial^2 u/\partial y^2$ values can occur for positive values of m, that is, when the pressure decreases throughout (the flow accelerates always). Turning points are excluded for the function $u(y)$.

The exact results of boundary layer theory for the special type $u_\delta \sim x^m$ in continuously accelerated ($m > 0$) or continuously retarded ($m < 0$) flows have given us insight into the character of the solutions of a more or less general nature, a fact that later will turn out useful in the development of approximate solutions.

In the general case of a boundary layer with an arbitrary free-stream flow $u_\delta(x)$ we cannot expect the successive velocity profiles in x direction to be similar to one another. When generalizing eq. (2.27) to the form

$$u_\delta(x) \sim x^{m(x)} \tag{2.47}$$

with an exponent m depending on x, we can expect that the velocity profiles, in a first approximation, develop according to the change of m with x. It should still be possible (in a first approximation) to represent the possible shapes of velocity profiles by the class of velocity profiles computed by Hartree [43], that is by the one-parametric family of curves of Fig. 2.2. This is one of the principal ideas of finding approximate solutions which will be discussed extensively in Chapter 3.

Whenever the exponent m changes rapidly with x, in particular, when m changes the algebraic sign with x (for example, when domains of pressure increase and decrease alternate periodically in the direction of flow) one has to expect a great variety of shapes for the velocity profiles which can only be coarsely approximated by a one-parametric family.

Until now there was for such problems only the very cumbersome approach of a direct solution of the system of partial differential equations (2.20), (2.21) by numerical methods (for example, series solutions or finite difference methods, see Görtler [38], or Witting [158].* A survey of the basic considerations of these methods is given in the book by H. Schlichting [111].

*A finite difference method by which the stability of the numerical computations can be constantly monitored has been published by Schönauer [116].

More recently, relatively simple and reliable approximate solutions have been developed for complicated problems of the type mentioned, using multiparametric (mostly two-parametric) families of functions for the velocity profiles in connection with integral conditions (which result from taking averages of Prandtl's boundary layer equation). Problems connected with and a critique of these approximate solutions are the main topic of Chapter 3.

Readers who are less interested in boundary layer problems for compressible flows and in questions concerning heat transfer may skip the following pages and go to Chapter 3.

2.3 The Energy Equation of the Temperature Boundary Layer

2.3.1 General remarks about the problem

This section is primarily concerned with a generalization of the exact foundations of boundary layer theory; this becomes necessary when variable material properties (such as density, viscosity, heat conductivity) have to be assumed. This general problem is presented, for example, at high flow velocities of a compressible medium (such as air) with and without additional heating or cooling of the fixed flow boundary. Changes of the velocity within and outside the boundary layer are here connected with changes in the temperature of the medium, and these in turn influence the material properties. Since *heat energy* is involved, the boundary layer problem can no longer be described by the conditions of force equilibrium and of continuity. In addition to the known equation of state for the density $\rho(p, T)$ for gases

$$\frac{p}{\rho} = RT, \tag{2.48}$$

where R = gas constant (cf. also Appendix II) and the empirical temperature law for the viscosity $\mu(T)$, eq. (1.7), we also have to consider *energy laws*. These additional equations, together with Prandtl's system of equations (2.20) and (2.21), form the exact basis for the theory of coupled flow and temperature boundary layers. In the following, we shall deal with the derivation of these additional relations and the possibilities of solving the system of simultaneous equations for this more general problem.

2.3.2 Thermodynamical relations for viscous flows

This section is concerned with energy considerations for the stationary flow of a compressible viscous medium (a gas). In order to display clearly some of the general connections we shall not limit ourselves to the specific boundary layer flow along a fixed wall for which Prandtl's boundary layer simplifications (Sections 1.6 and 1.7) hold. Our considerations are rather concerned with a filament of flow. For simplicity's sake, we assume an ideal gas as the flowing medium.

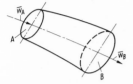

Fig. 2.3. Illustration of the energy balance for the unit mass flowing in a flow filament.

It is characteristic for a flow filament that the same mass traverses every cross section. The local cross section shall be denoted by F, the local average velocity in the filament denoted by $\bar{w}$, and the time by t. Limiting our interest to a segment of such a filament between the planes A and B (Fig. 2.3), we find that at A the mass

$$\bar{\rho}_A \bar{w}_A F_A \, dt = m_A \tag{2.49}$$

enters through that cross section, while at B the mass

$$\bar{\rho}_B \bar{w}_B F_B \, dt = m_B \tag{2.50}$$

leaves. Here $\bar{\rho}$ is an average over the cross section. Both masses are equal for reasons of continuity:

$$m_A = m_B = m. \tag{2.51}$$

We ask now for energy transfers that are possible within this mass m (chemical and atomic processes are excluded). It is obviously necessary to consider the following forms of energy:*

(a) Kinetic energy E_K (mechanical energy)
(b) Internal energy E_i (heat energy, energy of molecular motion)
(c) Mechanical work of the pressure force, E_p (mechanical energy)
(d) Mechanical work of the shear stresses, E_τ (mechanical energy).

---

*Strictly speaking, when listing the mechanical forms of energy we should also consider the potential energy of the medium. This form of energy plays an important role in hydraulic machines and more generally in flows with free surfaces, but not so with boundary layer flows caused by "forced convection." Such "free convection boundary layers" are not within the scope of this book.

The sum of the changes of all these energies has to be 0, or equal to the difference dE_q of the heat energy which is added or escapes by heat conduction between A and B. The effect of heat radiation is important only for very high temperatures (about $T > 1000°$ K) and shall be neglected against heat conduction within the framework of the subsequent investigations.

We shall now formulate the above-mentioned energies. To this end we introduce for velocity, pressure, density, temperature, and the work of the shear stresses values that have been averaged over the cross section of the filament of flow, namely, $\overline{w}$, $\overline{p}$, $\overline{T}$, and $\overline{w\tau}$.

(a) *Kinetic energy* E_K: by definition we have

$$E_K = m\frac{\overline{w}^2}{2}.$$ (2.52)

(b) *Internal energy* E_i: according to the laws of thermodynamics the energy of molecular motion is a function of the temperature T and for ideal gases is given by

$$E_i = mc_v\overline{T},$$ (2.53)

where c_v is the (generally constant) specific heat at constant volume. Denoting by c_p the (generally constant) specific heat at constant pressure and by

$$\kappa = \frac{c_p}{c_v},$$ (2.54)

$$R = c_p - c_v,$$ (2.55)

$$\frac{c_v}{R} = \frac{1}{\kappa - 1}$$ (2.56)

we may, by observing (2.48) write eq. (2.53) also in the form

$$E_i = m\frac{1}{\kappa - 1}\frac{\overline{p}}{\overline{\rho}}.$$ (2.57)

(c) *Mechanical work* of the *pressure* E_p: mechanical work is given by the product of force and the length of the path. Thus we have

$$E_p = \overline{p}F \cdot \overline{w}\, dt = \overline{\rho}\overline{w}F\, dt \cdot \frac{\overline{p}}{\overline{\rho}} = m\frac{\overline{p}}{\overline{\rho}}.$$ (2.58)

The quantity $1/\overline{\rho}$ is also called "specific volume."

(d) *Mechanical work of the shear stresses:* for the work of the shear stress one obtains

$$E_\tau = F \frac{\overline{w\tau}}{\overline{w}} \, \overline{w} \, dt = \bar{\rho} F \overline{w} \, dt \frac{\overline{w\tau}}{\bar{\rho}\overline{w}} = m \frac{\overline{w\tau}}{\bar{\rho}\overline{w}} . \tag{2.59}$$

In all forms of energy the mass m appears as a multiplier. It is therefore possible to formulate the energy balance for the unit mass by dividing all terms by m. Energies relating to the unit mass shall be denoted by the letter e. At the same time, we shall consider only energy changes. Then the balance of energy is given by

$$d\left(\frac{\overline{w}^2}{2}\right) + d\left(\frac{\bar{p}}{\bar{\rho}}\right) + d(e_i)$$
$$= d(e_\tau) + d(e_q) = \begin{cases} \text{energy having flown} \\ \text{in or out of the} \\ \text{surface of the filament.} \end{cases} \tag{2.60}$$

Here the quantity $d(e_q)$ is the contribution by heat conduction. The heat transfer by radiation, which can be easily computed according to known laws, has been neglected for the sake of simplicity.

In the case where there is no exchange of mechanical work due to shear stresses at the boundary of the filament, the work done by the shear stresses in the interior of the filament at the expense of the other mechanical energies appears entirely as positive heat. In this special case, we have $d(e_\tau) = 0$.

It is customary in thermodynamics to call the sum

$$\frac{p}{\rho} + e_i = i \tag{2.61}$$

the *enthalpy* or *heat content*. In the case of an ideal gas we have, moreover,

$$i = c_p T. \tag{2.62}$$

It is further customary to define the sum of kinetic energy and enthalpy

$$\frac{w^2}{2} + i = h \tag{2.63}$$

as the *total energy*. With use of eq. (2.63), the balance of energy (2.60) can also be written in the form

$$d\bar{h} = d(e_\tau) + d(e_q). \tag{2.64}$$

Equation (2.64) expresses the fact that the total energy (even in a viscous medium) can only change by heat transfer to or from the surrounding medium and by exchange of energy, e_τ, due to viscous forces.*

The following remarks are important as they pertain to the work done by the pressure forces p/ρ on the unit of mass. For gases the specific volume $1/\rho$ is a function of the pressure p. For an adiabatic change of state we have

$$p\left(\frac{1}{\rho}\right)^\kappa = \text{const.} \tag{2.65}$$

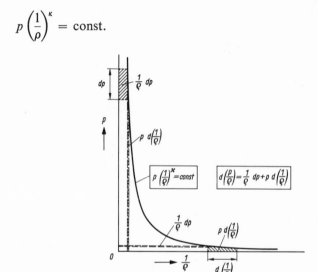

Fig. 2.4.   Relation between pressure p and specific volume $1/\rho$ of an ideal gas. For isentropic changes of state, $p(1/\rho)^\kappa = \text{const.}$ ($\kappa = 1.4$), as shown in the curve. For incompressible changes of state ($\kappa \to \infty$), $\mathrm{d}(p/\rho) = 1/\rho\ \mathrm{d}p$.

(Fig. 2.4). Using the differential notation for the balance of energy, we have therefore

$$\mathrm{d}\left(\frac{p}{\rho}\right) = \frac{1}{\rho}\mathrm{d}p + p\,\mathrm{d}\left(\frac{1}{\rho}\right). \tag{2.66}$$

We can see from eq. (2.66) that the work done by the pressure forces can manifest itself in two different ways: by change of pressure at constant specific volume $1/\rho$, that is, "*mechanical work,*" or by change of specific volume $1/\rho$ at constant pressure p, that is, by work due to compression or expansion. Figure 2.4 illustrates these changes of p/ρ for high pressure

*In deriving this balance of energies, we often find in the literature (see, e.g., Prandtl [95] that $\mathrm{d}(e_\tau) = 0$ is assumed; this is permissible only in special cases.

(practically incompressible media)* and also at very low pressure (large changes of the specific volume $1/\rho$ with the pressure p).

2.3.3 Derivation of the differential equation for the temperature boundary layer

2.3.3.1 Preliminary remarks. As in Fig. 2.1, we consider a control section dx, dy of unit width in a two-dimensional flow in the x, y domain. When formulating the energy balance for this control section we have to note that in addition to the mass flows ρu and ρv through this section there are also energy flows associated with them, such as "flow of enthalpy" and "flow of total energy." This case of energy transfer is called "convection." The mass flow satisfies the continuity condition (2.21). For the equation of motion we may use Prandtl's boundary layer simplifications (Section 1.7) for large Reynolds numbers, that is, for example, to neglect the pressure gradient $\partial p/\partial y$ perpendicular to the wall and to put $\partial u/\partial y \gg \partial u/\partial x$ and $\partial^2 u/\partial y^2 \gg \partial^2 u/\partial x^2$. Moreover, because of eq. (1.49), we may neglect the energy $v^2/2$ (per unit mass) in y direction against the kinetic energy $u^2/2$ in x direction in the definition (2.63) for the total energy h. On the basis of Prandtl's boundary layer simplifications we may therefore write for the total energy h:

$$h = \frac{u^2}{2} + i. \tag{2.67}$$

From our findings in Section 2.3.2 we may further conclude that large temperature gradients can be expected whenever the shear stresses and their work transformed into heat (dissipation) are large, viz., within the flow boundary layer. We may therefore assume that the thickness of the temperature boundary layer is of the same order of magnitude as the flow boundary layer as long as the heat exchange by conduction in y direction is not of a different order of magnitude than the exchange of momentum by viscosity, which is responsible for the increase of thickness of the flow boundary layer. We shall see that this physical situation is characterized by a dimensionless parameter

$$\text{Pr} = \frac{\mu c_p}{\lambda}. \tag{2.68}$$

The Prandtl number Pr can be interpreted as the ratio of the momentum exchange, which is proportional to μ, and the temperature exchange, which is proportional to λ/c_p. For a laminar boundary layer Pr is a pure

*Incompressibility can actually be realized only for $\kappa \to \infty$.

material property. For turbulent boundary layers the exchange of momentum and temperature is increased by the turbulent motions by about the same ratio. One defines a "turbulent Prandtl number" Pr_t by

$$Pr_t = \frac{(\mu + \mu_s)c_p}{\lambda + \lambda_s} = \frac{\mu_e c_p}{\lambda_e}, \tag{2.69}$$

where μ_s and λ_s are the "apparent" and μ_e, λ_e the "effective" values of viscosity and of heat conductivity [see eqs. (1.21) and (1.25)]. For large Reynolds numbers, which have to be assumed within the domain of validity of Prandtl's boundary layer theory, we have $\mu_s/\mu \gg 1$ and $\lambda_s/\lambda \gg 1$, and thus

$$Pr_t \approx \frac{\mu_s c_p}{\lambda_s}. \tag{2.70}$$

The turbulent Prandtl number of air has values in boundary layers between about 0.8 and 0.9, indicating but little change against the molecular Prandtl number $Pr = 0.72 \pm 0.02$ ($50° K < T < 1600° K$), even under strong influence of compressibility (see, e.g. [17]). The often-used simplification $Pr = Pr_t = 1$ holds even better for turbulent boundary layers in air than for laminar ones.*

We should note here that for heat transfer along a fixed wall ($y = 0$) we have to use the molecular heat conductivity λ in Fourier's equation of heat conduction (1.24), since for $y \to 0$ the turbulent exchange motion disappears and therefore $\lambda_s \to 0$.

If Pr or Pr_t is of unit magnitude, then the thicknesses of the boundary layers for flow and temperature are of the same magnitude. For large Reynolds numbers the temperature boundary layer is then very thin. It follows that the temperature gradient $\partial T/\partial y$ perpendicular to the wall will be much larger than the temperature gradient $\partial T/\partial x$ in direction of flow.† For $R_L \gg 1$, the boundary layer simplifications reduce, besides

$$\frac{\partial p}{\partial y} = 0, \quad \frac{\partial u}{\partial y} \gg \frac{\partial u}{\partial x}, \quad \frac{\partial^2 u}{\partial y^2} \gg \frac{\partial^2 u}{\partial x^2} \tag{2.71}$$

*In the laminar sublayer of turbulent boundary layers in liquids the molecular Prandtl number may be very much larger than unity (for water at 20 °C, $Pr \approx 7$; for motor oil at 20 °C, $Pr \approx 10^4$) or very much smaller than unity (liquid metals, such as mercury, have a Prandtl number at 20 °C of $Pr \approx 0.023$). The influence on the heat transfer caused by this effect can easily be accounted for (see Section 2.3.8).
†At least as long as the gradient of the wall temperature dT_w/dx, which can be prescribed, is small, as we assume at present (see also Section 5.3).

also to the following statements about the orders of magnitude:

$$\frac{\partial T}{\partial y} \gg \frac{\partial T}{\partial x}, \quad \frac{\partial^2 T}{\partial y^2} \gg \frac{\partial^2 T}{\partial x^2}. \tag{2.72}$$

When checking the energy balance for the control section dx, dy (Fig. 2.1) one has to use the additional boundary layer simplification (2.72) with some reservation and to check later to what extent this was permissible.

2.3.3.2 The energy balance. From our considerations of Section 2.3.2 we are already familiar with the types of energies that enter the analysis. These energies are the total energy, the work done by the shear stresses including dissipation, and the heat energy, which enters or leaves the system by conduction. Work done by the pressure does not appear explicitly in the balance for the total energy, because of the definition (2.61) for the enthalpy.

It is thus possible to write down the partial differential equation for the total energy based on the boundary layer simplifications (2.71) and (2.72) in a manner quite similar to the derivation of eq. (2.20). The energy amounts flowing into the control section are denoted by a positive sign.

The total energy that flows with the mass flow into the control section from the left per unit of time and width is given by

$$\rho u h \, dy. \tag{2.73}$$

On the right side the total energy

$$-\left(\rho u h + \frac{\partial(\rho u h)}{\partial x} dx\right) dy \tag{2.74}$$

leaves the control section. The change in total energy affected by the mass flow ρu in x direction is therefore given by

$$-\frac{\partial(\rho u h)}{\partial x} dx \, dy. \tag{2.75}$$

The change in total energy resulting from the mass flow ρv in y direction through the control section is correspondingly

$$-\frac{\partial(\rho v h)}{\partial y} dy \, dx. \tag{2.76}$$

The entire change of total energy resulting from "forced convection" can thus be written

$$-\left[\frac{\partial(\rho u h)}{\partial x} + \frac{\partial(\rho v h)}{\partial y}\right] dx\,dy. \tag{2.77}$$

After differentiation, expression (2.77) becomes

$$-\left\{\rho u\frac{\partial h}{\partial x} + \rho v\frac{\partial h}{\partial y} + h\left[\frac{\partial(\rho u)}{\partial x} + \frac{\partial(\rho v)}{\partial y}\right]\right\} dx\,dy. \tag{2.78}$$

The expression inside [ ] is the equation of continuity (2.21) multiplied by h; in other words, it vanishes identically.

Thus, the change in total energy by *forced convection* for the control section is given by

$$-\left(\rho u\frac{\partial h}{\partial x} + \rho v\frac{\partial h}{\partial y}\right) dx\,dy. \tag{2.79}$$

In order to write down the *work* done by the *shear stresses*, we refer to Fig. 1.1 and the remarks in Section 1.2. At the lower boundary of the control section the transfer of work done by the shear stress (assumed negative) is

$$-u\tau\,dx, \tag{2.80}$$

which amount is transferred (i.e., lost) to the neighboring section closer to the wall. At the upper boundary, the control section receives the (positive) work done by the shear stresses

$$\left[u\tau + \frac{\partial(u\tau)}{\partial y}\,dy\right] dx. \tag{2.81}$$

Thus, we have the (positive) difference:

$$\frac{\partial(u\tau)}{\partial y}\,dy\,dx, \tag{2.82}$$

which is the energy of the shear stresses that interacts with the other forms of energy. (The y component of the shear energy is negligible within the framework of the assumed boundary layer simplifications by Prandtl.) The important question what fraction of the shear energy (2.82) is being dissipated will be dealt with later in more detail.

According to eq. (2.64) we have to add in the differential equation to be established for the total energy a term expressing the *heat transfer*. Conductive heat transfer can take place only between regions of different temperatures, where the heat flows from the higher to the lower temperatures. Denoting by λ_e the effective heat conductivity given by eq. (1.25), we write for the heat energy that enters the control section across the lower boundary (per unit of time and width)

$$q \, dx = -\lambda_e \frac{\partial T}{\partial y} \, dx. \tag{2.83}$$

When $\partial T / \partial y$ is negative, the heat flow appears as an increase in energy in the control section and is therefore positive.

At the upper boundary of the control section, the amount of heat energy that leaves by conduction is

$$-\left(q + \frac{\partial q}{\partial y} \, dy\right) dx = \left[\lambda_e \frac{\partial T}{\partial y} + \frac{\partial}{\partial y}\left(\lambda_e \frac{\partial T}{\partial y}\right) dy\right] dx. \tag{2.84}$$

Per unit of time and width, the control section therefore loses the heat energy (the difference between entering and leaving heat energy)

$$\frac{\partial}{\partial y}\left(\lambda_e \frac{\partial T}{\partial y}\right) dy \, dx. \tag{2.85}$$

A temperature gradient in x direction similarly results in a change of heat energy in the control section, of the magnitude

$$\frac{\partial}{\partial x}\left(\lambda_e \frac{\partial T}{\partial x}\right) dx \, dy. \tag{2.86}$$

Under the assumption of the boundary layer simplification (2.72) this expression (2.86) may however be neglected against (2.85).

The sum of energy changes (2.79), (2.82), and (2.85) has to be zero. Thus, the partial differential equation for the total energy in two-dimensional flow is (after division by a common factor $dx \, dy$)

$$\rho u \frac{\partial h}{\partial x} + \rho v \frac{\partial h}{\partial y} = u \frac{\partial \tau}{\partial y} + \tau \frac{\partial u}{\partial y} + \frac{\partial}{\partial y}\left(\lambda_e \frac{\partial T}{\partial y}\right). \tag{2.87}$$

Multiplying Prandtl's boundary layer eq. (2.20) by u [by making use of the identity $u \, \partial u = \partial(u^2/2)$], we arrive at the additional partial differential equation for the mechanical energies

$$\rho u \frac{\partial\left(\frac{u^2}{2}\right)}{\partial x} + \rho v \frac{\partial\left(\frac{u^2}{2}\right)}{\partial y} = -u \frac{dp}{dx} + u \frac{\partial \tau}{\partial y}. \tag{2.88}$$

Subtraction of eq. (2.88) from (2.87) results, together with (2.67), in the equation for the enthalpy field:

$$\rho u \frac{\partial i}{\partial x} + \rho v \frac{\partial i}{\partial y} = u \frac{dp}{dx} + \tau \frac{\partial u}{\partial y} + \frac{\partial}{\partial y} \left(\lambda_e \frac{\partial T}{\partial y} \right).$$ (2.89)

Because

$$i = e_i + \frac{p}{\rho} = c_p T$$ (2.90)

holds for ideal gases, there follows the equation for the temperature field

$$c_p \left(\rho u \frac{\partial T}{\partial x} + \rho v \frac{\partial T}{\partial y} \right) = u \frac{dp}{dx} + \tau \frac{\partial u}{\partial y} + \frac{\partial}{\partial y} \left(\lambda_e \frac{\partial T}{\partial y} \right).$$ (2.91)

Differentiating the left-hand side of eq. (2.89) and using the definition (2.90) for the enthalpy i, together with relation (2.66), lead to

$$p \left(\frac{\partial u}{\partial x} + \frac{\partial v}{\partial y} \right) + \rho u \frac{\partial e_i}{\partial x} + \rho v \frac{\partial e_i}{\partial y} = \tau \frac{\partial u}{\partial y} + \frac{\partial}{\partial y} \left(\lambda_e \frac{\partial T}{\partial y} \right).$$ (2.92)

It appears useful here to indicate that for the mass-preserving system of Section 2.3.2, eq. (2.92) assumes the form

$$p \, d \left(\frac{1}{\rho} \right) + d(e_i) = \tau \, d \left(\frac{1}{\rho} \right) + d(e_q).$$ (2.93)

This expresses the fact that heat transferred to the unit of mass by conduction or dissipation [the terms on the right-hand side of eq. (2.93)] is used to increase the internal energy and to do expansion work.

The transfer of kinetic energy into heat energy as a result of work of the shear stresses does not change the balance for the total energy, but it cannot be used to restore the original state of motion (for example, the original kinetic energy). In eqs. (2.92) and (2.93) the term $(1/\rho) \, dp$, which represents that part of the potential energy of the pressure that can be transformed in kinetic energy (mechanical work), has to be missing. This fact expresses the irreversibility of the dissipation mechanism.

In a flow without viscosity (thus without frictional heat) and without heat flow to or from the medium, i.e., in an isentropic flow, as it exists in practice outside of the boundary layer, the two terms on the right-hand side of eq. (2.93) are zero. In that case only the internal energy e_i and the work done by expansion or compression, $pd(1/\rho)$, interact. Thus, we have in the case $d(e_q) = 0$, $\tau \, d(1/\rho) = 0$:

$$d(e_i) = -p \, d \left(\frac{1}{\rho} \right).$$ (2.94)

Substituting eq. (2.94) into (2.60), and considering eq. (2.66), one obtains

$$d \left(\frac{w^2}{2} \right) + \frac{1}{\rho} \, dp = 0,$$ (2.95)

which is Bernoulli's law for a compressible medium without friction. In integrated form, this law [which appears already in eq. (1.59)] becomes

$$\frac{w^2}{2} + \int \frac{1}{\rho}\,dp = \text{const.} \tag{2.96}$$

For constant density ρ, i.e., for incompressible flow, eq. (2.96) assumes the well-known form:

$$\rho\frac{w^2}{2} + p = \text{const.} \tag{2.97}$$

The following important remark is in order about the work done by the shear stresses (2.82): after carrying out the differentiation in expression (2.82), we obtained in eq. (2.87) terms that correspond to

$$\frac{\partial(u\tau)}{\partial y}\,dy\,dx = \left[u\frac{\partial \tau}{\partial y} + \tau\frac{\partial u}{\partial y}\right]dx\,dy. \tag{2.98}$$

The first term in the bracket,

$$u\frac{\partial \tau}{\partial y}\,dx\,dy, \tag{2.99}$$

obviously represents the mechanical energy due to the shear stresses which is taken away or (for positive values of $\partial \tau/\partial y$) added to the system. By the definition $\tau = \mu_e\,\partial u/\partial y$, the second term can also be written as

$$\tau\frac{\partial u}{\partial y}\,dx\,dy = \mu_e\left(\frac{\partial u}{\partial y}\right)^2 dx\,dy; \tag{2.100}$$

this has always a positive algebraic sign and represents that portion of the work done by the shear stresses which is transformed into heat, i.e., dissipation. It is thus clear that the term (2.98) includes both mechanical energy which is added or taken away from the flow by the shear stresses and energy transformed into heat (dissipation).

It should be noted that the two contributions (2.99) and (2.100) are not necessarily of equal size. If the product $u\tau$ does not happen to be zero at both the lower and upper boundaries of the control section (for example, when $u = 0$ or $\tau = 0$), then work done by the shear stresses is transferred to the neighboring control section or is received from there. A part or even all dissipated energy in a given control section may derive from a neighboring domain. An example for this case is given by the shear flow between two moving walls: owing to the constant shear stress in Couette flow [$\tau = \mu_e\,(\partial u/\partial y) = \text{const}$], the derivative $\partial \tau/\partial y$ is zero. Because of eq. (2.99), this shear flow does not lose any mechanical (kinetic) energy. However, energy is being dissipated throughout the entire shear flow since the term $\mu_e\,(\partial u/\partial y)^2$ does not vanish due to $\partial u/\partial y$ being constant. The mechanical equivalent for this dissipated energy is being produced by the movement of the walls with the relative velocity u against the shear stress $\tau = \tau_w = \text{const.}$

Equation (2.87) for the total energy can be simplified and made physically more understandable by a simple transformation. In the term for the heat conduction on the right-hand side of that equation, we replace the enthalpy i, as given by eq. (2.67), by the total energy h and introduce the molecular, or turbulent, Prandtl number Pr, or Pr_t, according to eq. (2.68) or (2.69), both of which are constant for a very large temperature domain (in the following we write Pr throughout*):

$$\frac{\partial}{\partial y}\left(\frac{\lambda}{\mu c_p}\mu\frac{\partial i}{\partial y}\right) = \frac{\partial}{\partial y}\left(\frac{1}{Pr}\mu\frac{\partial h}{\partial y}\right) - \frac{\partial}{\partial y}\left(\frac{1}{Pr}u\mu\frac{\partial u}{\partial y}\right)$$

$$= \frac{1}{Pr}\left[\frac{\partial}{\partial y}\left(\mu\frac{\partial h}{\partial y}\right) - \frac{\partial(u\tau)}{\partial y}\right]. \qquad (2.101)$$

Thus, eq. (2.87) can be written in the form

$$\rho u\frac{\partial h}{\partial x} + \rho v\frac{\partial h}{\partial y} = \frac{Pr - 1}{Pr}\frac{\partial(u\tau)}{\partial y} + \frac{1}{Pr}\frac{\partial}{\partial y}\left(\mu\frac{\partial h}{\partial y}\right) \qquad (2.102)$$

with the boundary conditions

$$y = 0: u = v = 0; \quad h(x, 0) = c_p T_w(x); \quad (u\tau)_w = 0, \qquad (2.103)$$

$$y \to \infty: u(x, \infty) \to u_\infty(x); \quad h(x, \infty) \to h_\infty; \quad \tau_\infty = 0; \quad (u\tau)_\infty = 0, \qquad (2.104)$$

or better (see Section 1.5)

$$y = \delta: u(x, \delta) = u_\delta(x); \quad h(x, \delta) = h_\delta; \quad \tau_\delta = 0; \quad (u\tau)_\delta = 0 \qquad (2.105)$$

($h_\delta = h_\infty$ = const. for isentropic external flow).

It is now suggested for the sake of simplicity to seek a solution of this partial differential equation of second order which is linear in h, first for the special case Pr = 1, which, as we mentioned earlier, is approximately satisfied, for example, by air with Pr = 0.72. In this case, because Pr − 1 = 0, the term that expresses the influence of the work of the shear stresses on the total energy vanishes. We are then left with the simple equation

$$\rho u\frac{\partial h}{\partial x} + \rho v\frac{\partial h}{\partial y} = \frac{\partial}{\partial y}\left(\mu\frac{\partial h}{\partial y}\right) \quad \text{for} \quad Pr = 1. \qquad (2.106)$$

---

*For constant c_p, the assumption Pr = const. implies also μ/λ = const. Independence of Pr from T results for laminar boundary layers also in the independence from the distance from the wall, y. In turbulent boundary layers, the turbulence mechanism assures the approximate constancy of the ratio μ_s/λ_s, and the Prandtl number Pr_t within the boundary layer (cf. Section 2.3.3.1). It is therefore permissible to bring Pr in eq. (2.101) to the front of the entire equation as a numerical factor.

It is worth noting that eq. (2.106) is valid for arbitrary, incompressible and compressible external flow $u_\delta(x)$ with and without heat transfer at the wall $y = 0$.

2.3.4 The coupling between flow and temperature boundary layers for an insulated wall and Pr = 1

The simple eq. (2.106) has the following particular solution in the case where there is no heat transfer at the wall, $y = 0$ (insulated wall, so-called "thermometer problem"):

$$h(x, y) = \text{const.} \tag{2.107}$$

From the boundary condition $y \to \infty$, $h \to h_\infty$ ($y = \delta$, $h = h_\delta = h_\infty$), it follows that the constant in eq. (2.107) has to be equal to h_δ. Under the assumptions

$$\text{Pr} = 1; \left(\frac{\partial T}{\partial y}\right)_{y=0} = \frac{1}{c_p}\left(\frac{\partial h}{\partial y}\right)_{y=0} = 0 \tag{2.108}$$

we find as the particular solution of (2.106)

$$h(x, y) = h_\delta(x) = \text{const.} \tag{2.109}$$

Equation (2.109) expresses the physically very significant fact that the total energy at every point x, y within the boundary layer is equal to the total energy of the isentropic external flow; this holds even for arbitrary external flow $u_\delta(x)$ and thus for arbitrary pressure gradients dp/dx.* With the definition (2.67) for h, it follows from eq. (2.109) that

$$\frac{u^2}{2} + c_p T = \frac{u_\delta^2}{2} + c_p T_\delta = c_p T_0 \tag{2.110}$$

(T_0 = temperature of the flow at $u = 0$, the so-called "rest temperature") or

$$\frac{T(x, y)}{T_\delta(x)} = 1 + \frac{u_\delta^2/2}{c_p T_\delta}\left[1 - \left(\frac{u}{u_\delta}\right)^2\right]. \tag{2.111}$$

The "temperature profile" T/T_δ, under the assumptions (2.108), is therefore uniquely coupled with the velocity profile u/u_δ.

The parameter $(u_\delta^2/2)c_p T_\delta$ of the isentropic external flow is the ratio of the kinetic energy to enthalpy (*Eckert number*); for ideal gases, according to known laws of gas dynamics, it is a function of the "local"

*The pressure gradient dp/dx appears neither in eq. (2.106) nor in eq. (2.102).

Mach number (see also Appendix II):

$$M_\delta(x) = \frac{u_\delta(x)}{a_\delta(x)} \tag{2.112}$$

[a_δ is the local (x-dependent) speed of sound which is known for a given flow problem]. We have

$$\frac{u_\delta^2/2}{c_p T_\delta} = \frac{\kappa - 1}{2} M_\delta^2 \tag{2.113}$$

where

$$\kappa = \frac{c_p}{c_v}† \tag{2.114}$$

[$\kappa = 1.4$ for air, constant within 1.5 percent in the domain 60° K$< T <$ 600° K]. Thus, we may write eq. (2.111) in the form

$$\frac{T(x, y)}{T_\delta(x)} = 1 + \frac{\kappa - 1}{2} M_\delta^2 \left[1 - \left(\frac{u}{u_\delta} \right)^2 \right] \tag{2.115}$$

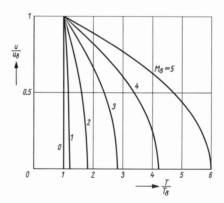

Fig. 2.5. Relation between the temperature profile $T(x, y)/T_\delta(x)$ and the velocity profile $u(x, y) / u_\delta(x)$ along a heat-insulated wall $[(\partial T/\partial y)_w = 0]$ at a Prandtl number Pr $= \mu c_p/\lambda = 1$ and for various Mach numbers of the isentropic free-stream flow.

Figure 2.5 represents $T/T_\delta(u/u_\delta, M_\delta)$ according to eq. (2.115) for some Mach numbers, while Fig. 2.6 shows the corresponding temperature

†For air at 273° K and 1 atm pressure, we have $\kappa = 1.4$. A very slight T dependence may be approximated by the formula $\kappa \approx 1.4222 - 0.8125 \, 10^{-4} \cdot T$, for 273° K $\leq$ $T \leq 1000°$ K. Generally, $\kappa = $ const is used.

profiles T/T_δ as a function of the dimensionless distance η from the wall, according to eq. (2.43), for the laminar boundary layer along a flat plate. For this representation, we assume for simplicity's sake the profiles u/u_δ to be those given by Blasius [4], i.e., for incompressible flows; this appears permissible for a schematic representation. From eq. (2.115) we may draw the following important conclusions:

(a) At the outer edge of the flow boundary layer which is characterized by $u/u_\delta = 1$ we have also $T/T_\delta = 1$. The temperature boundary layer extends therefore just as far as the flow boundary layer. If we assume a finite thickness δ_S for the flow boundary layer, within the framework of an approximate theory, then the thickness δ_T of the associated temperature boundary layer according to (2.115) is the same:

$$\frac{\delta_S}{\delta_T} = 1 \quad \text{for} \quad \text{Pr} = 1. \tag{2.116}$$

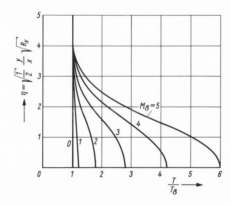

Fig. 2.6. Temperature profiles of Fig. 2.5 plotted as functions of the nondimensional distance from the wall, η, for the laminar boundary layer along the flat plate with u/u_δ at $M_\delta = 0$ according to Blasius [4], Pr = 1. The curves for $M_\delta > 0$ are only qualitative; this graph is thus only schematic and not suited for quantitative comparisons.

The temperature boundary layer therefore extends in y direction as far as the flow boundary layer. This subsequently justifies assumption (2.72).

(b) At the wall $y = 0$, $u = 0$ the "rest temperature" T_0

$$T_0 = T_\delta \left(1 + \frac{\kappa - 1}{2} M_\delta^2 \right) \tag{2.117}$$

results, which would also occur at an adiabatic stagnation point of the free-stream flow. It is therefore immaterial for the heating of the wall whether the flow was decelerated to velocity 0 by friction and adhesion or by stagnation.

For $M_\delta = 1$ we have from eq. (2.117), with $\kappa = 1.4$ for air, $T_0 = 1.2\,T_\delta$; for $M_\delta = 3.16$ ($M_\delta^2 = 10$) we have $T_0 = 3\,T_\delta$.

Assuming the (absolute) temperature T_δ of the free-stream flow to be $T_\delta = 300°$ K (27° C), the wall in the first case is heated to $T_0 = 360°$ K (87° C), in the second case to $T_0 = 900°$ K (627° C).

We note again that eqs. (2.115) through (2.117), which were derived under the assumption Pr $= 1$, are valid in boundary layers with arbitrary pressure gradients dp/dx and therefore for arbitrary free-stream flows $u_\delta(x)$ (however, only for insulated walls).

These equations are also valid, in good approximation, for the time average of the velocity in turbulent boundary layers. It is immaterial for the total energy h whether the dissipation of energy is caused by a direct action of the molecular viscosity or via the detour of turbulent exchange motions whose kinetic energy is finally (however, some with a certain time delay and at a different location) transformed into heat by molecular viscosity.

Aside from the particular solution discussed here in the special case of an insulated wall and Pr $= 1$ we are, of course, also interested in solutions of eq. (2.106) for Pr $= 1$ with heat exchange at the wall, as well as solutions of the general eq. (2.89) or (2.102) for Prandtl numbers other than 1 as they occur in fluids (water, with Pr ≈ 7 at 20° C; motor oil, with Pr $\approx 10^4$ at 20° C; in both cases there is strong temperature dependence of the Prandtl number). We shall deal with such solutions in the following sections.

2.3.5 Similar solutions for compressible flow with and without heat transfer and Pr $= 1$

No exact general solution is known of the simultaneous system of equations (2.20), (2.21), and (2.89) with the additional relations for the temperature and pressure dependence of the material properties ρ and μ. However, more recently, exact solutions for a few important special cases have been developed. Li and Nagamatsu [66] showed that, under the assumptions Pr $= 1$ and $\mu \sim T$, there exist "similar solutions" of the aforementioned system of equations for certain free-stream flows in the case of compressible flows with heat exchange, exactly as in the case of incompressible flow (Section 2.1.2). The velocity profiles u/u_δ and the profiles of the total energy $(h - h_\delta)/(h_w - h_\delta)$ are similar at every point x if a suitable measure for the distance from the wall y and the velocity u is chosen. The associated free-stream flows $u_\delta(x)$ are also of the type

$u_\delta(x) \sim x^m$, except that here m is defined differently than in Section 2.2.1. We refer to the extensive discussion on these papers in [111]. We merely point out here that the velocity profiles u/u_δ represent formally the same solution as in incompressible flow, with the only exception that the dimensionless distance from the wall, η, is defined differently from eq. (2.43) (Fig. 2.7). The profiles of the total energy appear to show little dependence

Figs. 2.7 and 2.8. Similar solutions for the simultaneous system of equations (2.20), (2.21), and (2.106) for compressible laminar boundary layer with heat transfer according to Li and Nagamatsu [66], for Pr = 1 and $\mu \sim T$. For definition of the shape parameter β, see [66].

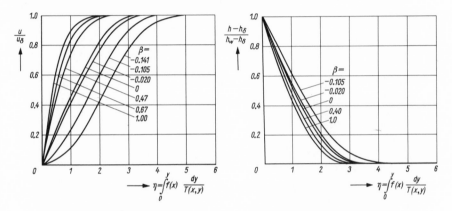

Fig. 2.7. Velocity profiles u/u_δ for specific free-stream flows $u_\delta(x) \sim x^m$ with $T_w/T_\delta = 2$.

Fig. 2.8. Profiles for the total energy $(h - h_\delta)/(h_w - h_\delta)$, associated with Fig. 2.7.

on the impressed pressure gradient dp/dx (Fig. 2.8), which appears plausible since the particular solution (2.115) for an insulated wall turns out to be independent of dp/dx. The influence of the pressure gradient dp/dx on the solution $h(x, y)$ of the general problem is apparently indirectly accounted for by a strong sensitivity of the velocity profile u/u_δ to the pressure gradient. This conclusion, drawn from the results of the special solution given by Li and Nagamatsu, is of similar importance for the later development of simple approximate solutions for the general boundary layer problem, as are the conclusions in Section 2.2.1.2 about the basic behavior of the flow boundary layer under pressure decrease and increase.

A special case of practical importance is the laminar compressible boundary layer without heat transfer in a Laval duct, a case treated by Geropp [34, 35]. His analysis is shown here in some detail. This particular case is one of the few in boundary layer theory where the simultaneous system of eqs. (2.20), (2.21), and (2.89) can be solved analytically in closed form. Based on the analyses by Li and Nagamatsu, Geropp, by observing the coupling relation (2.115) and assuming $\mu \sim T$ and $Pr = 1$, was able to derive one first-order ordinary differential equation each for the velocity distribution of the inviscid external flow and the velocity distribution within the boundary layer.

The solution of the ordinary differential equation for the velocity distribution of the inviscid external flow is

$$\int_{M_A^*}^{M^*} \frac{dM^*}{\left[1 - \dfrac{\kappa - 1}{\kappa + 1} M^{*2}\right]^{\frac{2\kappa-1}{\kappa-1}} M^{*2}} = \frac{x}{L} \tag{2.117a}$$

where M_A^* is the critical Mach number at the orifice of the duct, $x = 0$, and L the length of the duct.

The integral in eq. (2.117a) can be evaluated in closed form in those cases where $2(2\kappa - 1)/(\kappa - 1)$ is an integer (e.g., for $\kappa = 1.4$). For air with $\kappa = 1.4$, the velocity distribution of the external flow is

$$-\frac{1}{M^*(1 - \frac{1}{6}M^{*2})^{3.5}}$$

$$+ \frac{1}{6}\left[\frac{8M^*}{7(1 - \frac{1}{6}M^{*2})^{3.5}} + \frac{48M^*}{35(1 - \frac{1}{6}M^{*2})^{2.5}} + \frac{64}{35}\frac{M^*}{(1 - \frac{1}{6}M^{*2})^{1.5}}\right.$$

$$\left.+ \frac{128}{35}\frac{M^*}{(1 - \frac{1}{6}M^{*2})^{0.5}}\right] = \frac{x}{L} + B = X. \tag{2.117b}$$

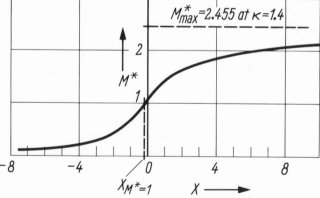

Fig. 2.9. Distribution of $M^*(X)$ in the duct.

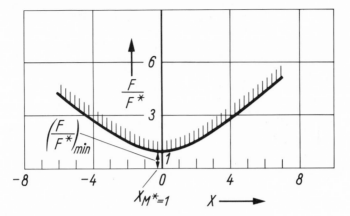

Fig. 2.10.   Contour of the duct.

Here B is a constant which is determined from the value of M_A^* at $X = 0$ with the aid of eq. (2.117b). The M* distribution that results from eq. (2.117b) (Fig. 2.9) was interpreted by Geropp [34] as the velocity distribution in a Laval duct, and he computed the corresponding contour using the one-dimensional flow theory (Fig. 2.10).

The solution of the ordinary differential equation for the velocity distribution (within the boundary layer) is:

$$
y = \frac{L\left(\dfrac{\kappa + 1}{2}\right)^{1/4}}{\sqrt{R_0}} \cdot \frac{1}{M^*\left[1 - \dfrac{\kappa - 1}{\kappa + 1}M^{*2}\right]^{\frac{1}{\kappa - 1}}}
$$

$$
\times \left\{ \sqrt{2}\tanh^{-1}\left(\frac{\sqrt{2 + \dfrac{u}{u_\delta}}}{\sqrt{3}}\right) - \sqrt{2}\tanh^{-1}\sqrt{\frac{2}{3}} \right.
$$

$$
\left. + \frac{(\kappa - 1)M^{*2}}{(\kappa + 1) - (\kappa - 1)M^{*2}}\left[\frac{2}{\sqrt{3}} - \sqrt{\frac{2}{3}}\left(1 - \frac{u}{u_\delta}\right)\sqrt{2 + \frac{u}{u_\delta}}\right] \right\}
$$

$$
(2.117c)
$$

where $R_0 = a_0 L/\nu_0$, a_0 is the velocity of sound, and ν_0 the kinematic viscosity at rest.

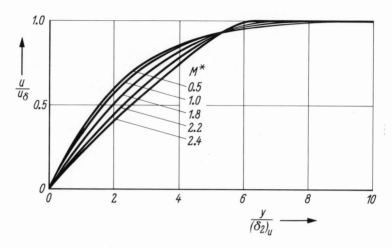

Fig. 2.11. Velocity profiles at different positions within the duct.

The various velocity profiles associated with the positions $X(M^*)$ of the Laval duct are displayed in Fig. 2.11. All other quantities of interest in the boundary layer may then be represented in closed form, using eq. (2.117c); thus, the displacement thickness, for example, becomes:

$$
\frac{\delta_1}{L} = \int_0^\infty \left(1 - \frac{\rho u}{\rho_\delta u_\delta}\right)\frac{dy}{L} = \frac{\left(\dfrac{\kappa+1}{2}\right)^{1/2}}{\sqrt{R_0}} \; \frac{1}{M^*\left(1 - \dfrac{\kappa-1}{\kappa+1}M^{*2}\right)^{\frac{1}{\kappa-1}}}
$$

$$
\times \left[3\sqrt{2} - 2\sqrt{3} + \frac{2}{\sqrt{3}}\,\frac{(\kappa-1)M^{*2}}{(\kappa+1) - (\kappa-1)M^{*2}}\right] \qquad (2.117d)
$$

The development of the displacement thickness along the contour of the duct is shown in Fig. 2.12. One recognizes that the displacement thickness decreases first in the subsonic section of the duct, reaches a minimum shortly before the narrowest cross section of the duct, and then increases rapidly in the supersonic section. According to Geropp [34, 35], these characteristic changes of δ_1 may be explained as follows: In the subsonic part of the flow, where the density barely changes, kinetic energy is added to the boundary layer by the rapidly accelerating free-stream flow which results in a decrease of the displacement thickness. In the subsequent supersonic part, the density in the boundary layer decreases rapidly as a

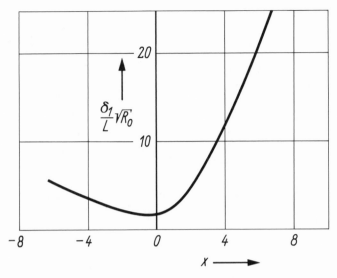

Fig. 2.12. Displacement thickness as a function of the position in the duct.

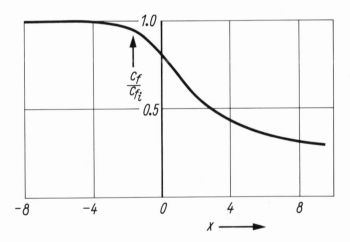

Fig. 2.13. Local coefficient of friction in the duct.

result of the rapid expansion of the free-stream flow and the great amount of frictional heat.

The kinetic energy of the boundary layer therefore decreases as well. As a consequence, the displacement thickness increases rapidly. The local coefficient of friction can also be written down explicitly for this flow in the duct of a given contour:

$$c_f = \frac{2\tau_w}{\rho_\delta u_\delta^2} = \frac{4}{\sqrt{3}} \frac{\left(\dfrac{\kappa + 1}{2}\right)^{1/4}}{\sqrt{R_0}} \left[1 - \frac{\kappa - 1}{\kappa + 1} M^{*2} \right] \tag{2.117e}$$

Figure 2.13 shows a plot of the local coefficient of friction c_f divided by the same quantity at the orifice of the duct, $c_{f_i}(M^* \approx 0)$, as a function of the distance along the wall from the orifice of the duct. It turns out that c_f/c_{f_i} decreases as the Mach number increases, again a consequence of the low density in the boundary layer.

In the following section we shall deal with a different exact solution of eq. (2.89), where we drop the assumption $\text{Pr} = 1$ but assume instead a solution with zero pressure gradient, $dp/dx = 0$ (for example, along a flat plate).

2.3.6 The coupling law between flow and temperature boundary layer in flows without pressure and temperature gradients in the direction of flow, at arbitrary Prandtl numbers

When discussing the solutions of the simultaneous system of eqs. (2.20), (2.21), and (2.106) in the two preceding sections, we assumed the Prandtl number to be equal to 1 in order to reduce mathematical difficulties. A physical reason for that assumption was given in Section 2.3.3.1. We are, however, interested in the error introduced by this simplifying assumption. Moreover, the problems associated with the temperature boundary layer and heat transfer are important in the fluid mechanics of liquid materials with very large Prandtl numbers, such as water and oil.

Van Driest [17] was able to show that the influence of the Prandtl number on the connection between flow and temperature layer can be accurately determined if one limits the problem to flows without pressure gradient ($dp/dx = 0$) and negligible temperature gradients in the direction of flow, as for example along a flat plate.* He used certain mathematical devices of older papers, notably those by Hantzsche and Wendt [42] and Crocco [10] which deal with the same problem.

The results of van Driest's research carry over to turbulent boundary layers as well and are therefore particularly suitable for the approximation theory which is to be developed in Chapter 3. The methods and results of this theory are here briefly illustrated.

*A different, simpler solution of this problem has been developed by Spalding [123] and by Kestin and Richardson [57]; however, at present this applies only to incompressible flow. The method of van Driest is preferred here because the compressibility effect is also accounted for.

Like Crocco [10], van Driest starts out from the system of eqs. (2.20), (2.21), and (2.89) and uses as the independent variables x and u, instead of x and y; i.e., he uses the transformation

$$x = x, \quad u = u(x, y). \tag{2.118}$$

The Prandtl number (Pr or Pr_t) is assumed constant but may have arbitrary values different from 1.

If for $dp/dx = 0$ and $\partial i/\partial x = 0$ we introduce two dimensionless quantities for shear and enthalpy

$$\tau^* = 2 \frac{\tau}{\rho_\delta u_\delta^2} \sqrt{R_x}, \quad R_x = \frac{\rho_\delta u_\delta x}{\mu_\delta}, \tag{2.119}$$

$$i^* = \frac{i}{i_\delta}, \tag{2.120}$$

the system (2.20), (2.21), and (2.89), transformed by eq. (2.118), becomes after elementary calculations a system of two ordinary differential equations (′ means derivative with respect to u/u_δ):

$$\tau^* \tau^{*\prime\prime} + 2 \frac{\mu}{\mu_\delta} \frac{\rho}{\rho_\delta} \frac{u}{u_\delta} = 0, \tag{2.121}$$

$$\left(\frac{i^{*\prime}}{Pr} \right)' + (1 - Pr) \frac{\tau'}{\tau} \frac{i^{*\prime}}{Pr} + \frac{u_\delta^2}{i_\delta} = 0. \tag{2.122}$$

For turbulent boundary layers, the Prandtl number Pr has to be replaced by Pr_t in these equations as well as in eqs. (2.125) through (2.131).

Because of eq. (2.119) it was possible to replace in eq. (2.122) $\tau^{*\prime}/\tau^*$ by τ'/τ (without *). The boundary conditions for eqs. (2.121) and (2.122) are

$$\frac{u}{u_\delta} = 0: \quad i^* = i_w^*, \quad \left[\left(\frac{\partial \tau}{\partial y} \right)_{y=0} = \frac{dp}{dx} = 0 \right], \tag{2.123}$$

$$\frac{u}{u_\delta} = 1: \quad i^* = 1, \quad (\tau = \tau_\delta = 0). \tag{2.124}$$

Equation (2.122) is linear in i^*, and, under the assumption that the shear stress distribution $\tau^*(u/u_\delta)$ is known, two successive integrations subject to the boundary conditions (2.123), (2.124) lead immediately to the general solution

$$i^* \left(\frac{u}{u_\delta} \right) = \frac{i}{i_\delta} \left(\frac{u}{u_\delta} \right) = 1 + \frac{i_e - i_w}{i_\delta} (f_1 - 1) + r \frac{u_\delta^2/2}{i_\delta} (1 - f_2). \tag{2.125}$$

Here the following abbreviations and definitions were used:

$$f_1 = \frac{\Sigma}{s};$$ (2.126)

$$f_2 = 2\frac{P}{r};$$ (2.127)

$$\Sigma = \text{Pr} \int_0^{u/u_\delta} \left(\frac{\tau}{\tau_w}\right)^{\text{Pr}-1} d\left(\frac{u}{u_\delta}\right);$$ (2.128)

$$P = \text{Pr} \int_0^{u/u_\delta} \left(\frac{\tau}{\tau_w}\right)^{\text{Pr}-1} \left[\int_0^{u/u_\delta} \left(\frac{\tau}{\tau_w}\right)^{1-\text{Pr}} d\left(\frac{u}{u_\delta}\right)\right] d\left(\frac{u}{u_\delta}\right);$$ (2.129)

$$(\Sigma)_0^1 = s = \text{Reynolds analogy factor}$$ (2.130)

$$2(P)_0^1 = r = \text{recovery factor}$$ (2.131)

$$i_e = i_\delta \left(1 + r\frac{u_\delta^2/2}{i_\delta}\right).$$ (2.132)

Since the shear distribution τ/τ_w is still unknown in eqs. (2.126) through (2.131), the expression for the solution (2.125) seems to be of little use. However, eqs. (2.128) and (2.129) show that for Prandtl numbers close to 1 the ratio $(\tau/\tau_w)^{\text{Pr}-1}$ is generally near unity, so that an estimate $(\tau/\tau_w)^{(0)}$ for an approximate solution $i*^{(1)}$ of eq. (2.125) should suffice. With $i*^{(1)}$, one may find an improved solution $(\tau/\tau_w)^{(1)}$ from eq. (2.121), and the complete solution of the system (2.121), (2.122) can be improved iteratively by using eq. (2.125).

Even when the Prandtl number differs widely from 1, convergence of the iterative method can be expected, as Hantzsche and Wendt [42] have shown. In any case one finds, for arbitrary Prandtl number and arbitrary shear distribution, from eqs. (2.126) through (2.131)

$$\frac{u}{u_\delta} \to 1: f_1 \to 1, \quad f_2 \to 1.$$ (2.133)

In Section 2.3.7 we shall discuss the conclusions that can be drawn for the thickness of the temperature boundary layer.

The following remarks are once more confined to air as the flowing medium, and the gas dynamic relations for ideal gases are assumed to hold. One has

$$i = c_p T;$$ (2.134)

$$\frac{u_\delta^2/2}{c_p T_\delta} = \frac{\kappa - 1}{2} M_\delta^2; \tag{2.135}$$

$$T_e = T_\delta \left(1 + r \frac{\kappa - 1}{2} M_\delta^2\right). \tag{2.136}$$

T_e is the "recovery temperature" and eq. (2.125) may be written as follows:

$$\frac{T}{T_\delta} = 1 + \frac{T_e - T_w}{T_\delta} (f_1 - 1) + r \frac{\kappa - 1}{2} M_\delta^2 (1 - f_2). \tag{2.137}$$

For a discussion of this solution, the Prandtl number is assumed to be $Pr = 1$, for the moment. Then it follows from eqs. (2.126) through (2.132) that

$$f_1 = \frac{u}{u_\delta}, \tag{2.138}$$

$$f_2 = \left(\frac{u}{u_\delta}\right)^2, \tag{2.139}$$

$$r = s = 1, \tag{2.140}$$

$$T_e = T_0, \tag{2.141}$$

and eq. (2.137) becomes

$$\frac{T}{T_\delta} = 1 + \frac{T_0 - T_w}{T_\delta}\left(\frac{u}{u_\delta} - 1\right) + \frac{\kappa - 1}{2} M_\delta^2 \left[1 - \left(\frac{u}{u_\delta}\right)^2\right], \tag{2.142}$$

where

$$T_0 = T_\delta \left(1 + \frac{\kappa - 1}{2} M_\delta^2\right) \tag{2.143}$$

is the *adiabatic stagnation temperature*.

A check of the boundary conditions shows that for $u/u_\delta = 1$ also $T/T_\delta = 1$ and that for $u/u_\delta = 0$ the temperature T is equal to the wall temperature T_w, which can be chosen arbitrarily. In the case of no heat transfer, characterized by $T_w = T_0$, eq. (2.142) becomes eq. (2.115), as expected.

If one introduces into eq. (2.142) the total energy $h = u^2/2 + c_p T$, one finds, with $h_w = c_p T_w$, (because of $u = 0$ for $y = 0$)

$$\frac{h - h_\delta}{h_w - h_\delta} = 1 - \frac{u}{u_\delta}. \tag{2.144}$$

For very small Mach numbers, $M_\delta \to 0$, we have, because of eq. (2.113), $u_\delta^2/2 \ll c_p T_\delta$, and thus

$$h_{M_\delta \to 0} \to c_p T. \tag{2.145}$$

Thus, for $M_\delta \to 0$ and $Pr = 1$, eq. (2.144) becomes

$$\frac{T - T_w}{T_\delta - T_w} = \frac{u}{u_\delta}. \tag{2.146}$$

If the temperature profile is defined by $(T - T_w)/(T_\delta - T_w)$, it becomes congruent with the velocity profile u/u_δ for incompressible flows with the heat transfer $(T_w = T_\delta)$, both in laminar and turbulent boundary layers.

The heat transfer is obviously controlled by the coefficient $(T_0 - T_w)/T_\delta$ in eq. (2.142), or by the coefficient $(T_e - T_w)/T_\delta$ in eq. (2.137), i.e., the temperature difference between the wall temperature and the recovery temperature divided by T_δ. In Section 2.3.8 we shall deal with this question extensively when we determine the heat transfer.

The aim of van Driest's [17] theory was to determine the influence of the Prandtl number on the solution for the temperature profile T/T_δ. The question is answered by comparing the solution (2.142) for $Pr = 1$ with the general solution (2.137) for ideal gases (or air) and arbitrary (but constant) Prandtl number. Equation (2.137) differs from eq. (2.142) by the fact that the values s and r from eqs. (2.130) and (2.131) are different from 1 and depend besides on Pr itself, as well as upon the Mach number through the term τ/τ_w and upon the heat transfer. Another fundamental difference between eqs. (2.137) and (2.142) consists in the replacement of u/u_δ and $(u/u_\delta)^2$ in eq. (2.137) by the functions

$$f_1 \left(\frac{u}{u_\delta}, \frac{\tau}{\tau_w}, Pr \right) \quad \text{and} \quad f_2 \left(\frac{u}{u_\delta}, \frac{\tau}{\tau_w}, Pr \right)$$

These functions are, like s and r, not only dependent on Pr but also upon the Mach number and heat transfer (because of the term τ/τ_w). Limiting our considerations to the flowing medium of air, where $Pr = 0.72 = const$, the small exponent $Pr - 1 = -0.28$ has the effect that the distribution of the shear stresses τ/τ_w has very little influence on f_1 and f_2 (and thus, also on s and r). It is, therefore, permissible to use an approximation for τ/τ_w, for example, $\tau(u/u_\delta)/\tau_w$ for the velocity profile of the flat plate in incompressible flow. In the case of a laminar boundary layer $\tau(u/u_\delta)/\tau_w$ has, according to Blasius [4], the typical characteristics displayed in

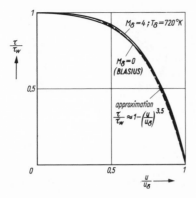

Fig. 2.14.   Shear $\tau(u/u_\delta)/\tau_w$ for the laminar velocity profile along a flat plate

—— according to Blasius [4]
- - - approximation $\tau/\tau_w = 1 - (u/u_\delta)^{3.5}$ $\Big\}M_\delta = 0$
—— according to van Driest [17]; $M_\delta = 4$, $T = 222°$K.

Fig. 2.14. In this case, the excellent approximation holds:

$$\frac{\tau}{\tau_w}\left(\frac{u}{u_\delta}\right) \approx 1 - \left(\frac{u}{u_\delta}\right)^{3.5}.$$ (2.147)

Proceeding according to van Driest [17], we obtain by iterative solution of the system (2.121), (2.122) for $M_\delta = 4$, $T_\delta = 222°$ K ($= 400°$ Rankine) the shear stress distribution shown in the upper solid curve in Fig. 2.14. It can be seen that the influence of the Mach number on $\tau(u/u_\delta)/\tau_w$ is very small, at least in the domain $0 < M_\delta < 4$. It is therefore permissible to use the approximation (2.147) for $\tau(u/u_\delta)/\tau_w$ in the case of a laminar boundary layer and Prandtl numbers in the order of magnitude of unity.

In Figs. 2.15a and b we have, based on the approximation (2.147), plotted the functions $f_1(u/u_\delta)$ and $f_2(u/u_\delta)$ with Pr as parameter (Pr $= 0.5$ to 10).* One finds that for $0.7 < $ Pr < 1 it is possible to put

$$f_1 \approx \frac{u}{u_\delta},$$ (2.148)

$$f_2 \approx \left(\frac{u}{u_\delta}\right)^2.$$ (2.149)

*Prandtl numbers very much larger or smaller than 1 essentially occur only with incompressible media (water, oil, liquid metals), for which certain simplifications of the theory are possible (see, for example Eckert [19], pp. 93–98).

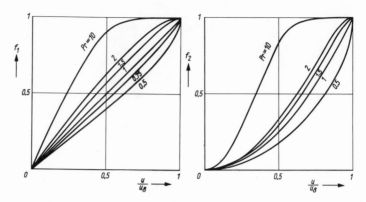

Fig. 2.15.   The functions f_1 and f_2 from eqs. (2.126) and (2.127) plotted against u/u_δ with the Prandtl number as parameter (Pr between 0.5 and 10).

The values of s and r, assuming (2.147) holds, are found for *laminar* boundary layer and Pr = 0.72 to be

$$s = 0.80, \tag{2.150}$$

$$r = 0.85. \tag{2.151}$$

Using the approximation $\tau/\tau_w \approx 1 - y/\delta$ and the Prandtl number $\mathrm{Pr}_t = 0.86$, averaged over the boundary layer thickness δ, one finds, according to van Driest [17], for the *turbulent* boundary layer the values

$$s = 0.82, \tag{2.152}$$

$$r = 0.88. \tag{2.153}$$

A more precise calculation by van Driest shows a minor decrease of s and r with increasing Mach number (of about 1 percent at $\mathrm{M}_\delta = 5$). For air with Pr = 0.72, eq. (2.137) may be simplified to

$$\frac{T}{T_\delta} = 1 + \frac{T_e - T_w}{T_\delta}\left(\frac{u}{u_\delta} - 1\right) + r\frac{\kappa - 1}{2}\mathrm{M}_\delta^2\left[1 - \left(\frac{u}{u_\delta}\right)^2\right]. \tag{2.154}$$

The fact that the Prandtl number of air is not equal to 1 means that in eq. (2.154) there is essentially a change in the numerical factor r, and the temperature difference $T_e - T_w$ occurs in place of $T_0 - T_w$. Analogous to eq. (2.143), we find, because of eqs. (2.132), (2.134) and (2.135):

$$T_e = T_\delta\left(1 + r\frac{\kappa - 1}{2}\mathrm{M}_\delta^2\right). \tag{2.155}$$

Equation (2.154) can now be written in the simpler form

$$\frac{T}{T_\delta} = a + b\frac{u}{u_\delta} + c\left(\frac{u}{u_\delta}\right)^2, \tag{2.156}$$

where the coefficients a, b, c are defined as follows

$$a = \frac{T_w}{T_\delta} = 1 + r\frac{\kappa - 1}{2}M_\delta^2 - \frac{T_e - T_w}{T_\delta}$$

$$= 1 + r\frac{\kappa - 1}{2}M_\delta^2(1 - \Theta), \tag{2.157}$$

$$b = \frac{T_e - T_w}{T_\delta} = \Theta\, r\frac{\kappa - 1}{2}M_\delta^2, \tag{2.158}$$

$$c = -r\frac{\kappa - 1}{2}M_\delta^2. \tag{2.159}$$

Here

$$\Theta = \frac{b}{r\dfrac{\kappa - 1}{2}M_\delta^2} = \frac{b}{r\dfrac{u_\delta^2/2}{c_p T_\delta}} = \frac{c_p(T_e - T_w)}{r u_\delta^2/2} = \frac{T_e - T_w}{T_e - T_\delta} \tag{2.160}$$

is the ratio between the heat energy transferred to the wall and the kinetic energy and may be introduced as the *heat transfer parameter*. The relations for the heat transfer (Sections 2.3.8 and 5.3) are greatly simplified if that parameter is used.

Positive values of Θ ($T_w < T_e$) mean cooling of the wall below the recovery temperature T_e which would obtain in the case of an insulated wall. Negative values of Θ ($T_w > T_e$) correspond to a heating of the wall above the recovery temperature T_e. For Mach numbers $M_\delta > 1$, generally only positive values of the parameter Θ are meaningful in practice, since according to eq. (2.117) it is necessary to counteract the aerodynamic heating of the wall by cooling. It is possible to limit the considerations, in general, to values between $\Theta = 0$ (insulated wall) and $\Theta = 1$ ($T_w = T_\delta$).

For incompressible boundary layers ($M_\delta = 0$), the coefficient b, defined in eq. (2.158), may be used as the parameter of heat transfer.

We emphasize that eqs. (2.156) and (2.157) through (2.160) were derived under the assumption that the pressure gradient dp/dx is equal to 0 in the free-stream flow but that the heat transfer is not restricted. Moreover, it should be remembered that eq. (2.115) is valid only for an insulated wall, but for any arbitrary pressure gradients dp/dx. We also point out that, according to the theoretically exact investigations by Li and Nagamatsu

[66] (Figs. 2.7 and 2.8) for similar solutions of compressible boundary layers with heat transfer and nonzero pressure gradients, the influence of the pressure gradient on the solution is generally small [only the temperature gradient in the vicinity of the wall $(\partial T/\partial y)_{y=0}$ is influenced].*

It therefore appears permissible to consider eq. (2.156) as a good approximation for the temperature distribution in laminar and turbulent boundary layers of a compressible medium, with and without heat transfer, for arbitrary pressure gradients dp/dx in the free-stream flow as long as the Prandtl number is confined approximately to the domain $0.7 < \text{Pr} < 1$.

Considerable influence on the temperature profile may be exerted by the temperature gradient dT_w/dx. Equation (2.156) is therefore only valid for small values of dT_w/dx. How to correct in special cases for the influence of dp/dx and of dT_w/dx on the temperature profile T/T_δ will be discussed more closely in Section 5.3.

Equation (2.156) is of extraordinary significance, especially for the subsequent development of approximation theories, because it expresses a unique relation between the temperature profile T/T_δ and the velocity profile u/u_δ, as do eqs. (2.111) and (2.115). The Mach number M_δ and the heat transfer parameter Θ in this coupling relation are considered parameters which, for each problem, are determined by the given functions $u_\delta(x)$, T_0, $T_w(x)$. Given the velocity profile, we obtain immediately the temperature profile T/T_δ by eq. (2.156) and, because of the boundary layer simplification $\partial p/\partial y = 0$, $p = p(x) = p_\delta(x)$, from the equation of state for gases, $\rho \sim 1/T$, also the density profile

$$\frac{\rho}{\rho_\delta} = \frac{T_\delta}{T}. \tag{2.161}$$

One may therefore consider the problem of the temperature boundary layer and its influence on density and viscosity solved, at least for the purposes of an approximation theory, for both laminar and turbulent boundary layers. For necessary improvements the extended theory in Section 5.3 is available.

The question of the heat transfer at the wall $y = 0$ is answered directly by eq. (1.24), which is the definition for heat transfer by conduction, provided the equation is applied at the wall $y = 0$ with $\lambda_e = \lambda$. Since the flowing medium adheres to the wall, heat can only be transferred there

---

*This is of secondary importance for the computation of the flow boundary layer but, as will be shown later, may prove significant (Section 5.3, Fig. 6.50) for the computation of the heat transfer.

by conduction. Heat transfer by convection does not take place up to a certain distance from the wall, but heat transfer by conduction at the wall is indirectly very strongly influenced by the fact that the temperature gradient $(\partial T/\partial y)_{y=0}$ in eq. (1.24) may be much larger with convection than without. This indirect influence of convection may exceed the effects of molecular conductivity.

In Section 2.3.8 we provide formulas necessary for the practical computation of heat transfer when the flow and temperature boundary layers are known (cf. also Section 5.3). But first some remarks are in order concerning the thickness of the temperature boundary layer when the Prandtl number differs from 1.

2.3.7 The thickness of the temperature boundary layer

In Chapter 1 we used the concept of boundary layer thickness to characterize the extent of the domain in which the viscosity is "practically" noticeable. In a rigorous sense, the influence of viscosity extends to infinitely large distances from the wall. A similar statement can be made about the temperature boundary layer: theoretically it extends out to infinity; in practice it is of finite thickness.

In the case where the Prandtl number $\mathrm{Pr} = 1$, we concluded already from eq. (2.115) that at the distance $y = \delta_\mathrm{S}$, where u/u_δ is practically equal to 1, T/T_δ is practically also equal to 1, and thus the thickness δ_T of the temperature boundary layer equals the thickness δ_S of the flow boundary layer [eq. (2.116)].

Information about the order of magnitude of the extent of the temperature boundary layer in comparison with the flow boundary layer for $\mathrm{Pr} \neq 1$ is easily obtained from eqs. (2.126) through (2.131) together with Figs. 2.15a, b and 2.16a, b. It is assumed that the temperature boundary layer has its outer edge where the functions f_1 and f_2 have practically reached the value unity. Figs. 2.15a and b show that this is the case for Prandtl numbers $\mathrm{Pr} > 1$ when $u/u_\delta < 1$. It is even more apparent than in Figs. 2.15a and b when f_1 and f_2 are plotted as functions of the distance from the wall η [eq. (2.43)] (see Figs. 2.16a and b). These graphs are valid for the laminar boundary layer along a flat plate with $u/u_\delta(\eta)$ according to Blasius [4] (cf. also Fig. 2.2, curve for $m = 0$).

The temperature boundary layer for $\mathrm{Pr} > 1$ is therefore thinner than the flow boundary layer (for water of 20° C with $\mathrm{Pr} = 7.03$, we have approximately $\delta_\mathrm{S}/\delta_\mathrm{T} \approx 2.65$, and for oil at 60° C with $\mathrm{Pr} = 1000$ we have $\delta_\mathrm{S}/\delta_\mathrm{T} \approx 10$). For air with $\mathrm{Pr} = 0.72$ the temperature boundary layer is slightly thicker than the flow boundary layer. We have according

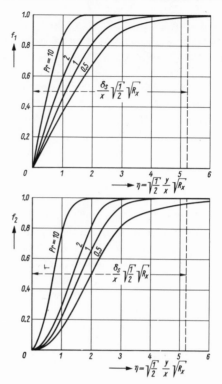

Fig. 2.16. The functions f_1 and f_2 of Figs. 2.15 plotted against the non-dimensional distance from the wall, η, for the laminar velocity profile $f'(\eta) = u/u_\delta$ of the flat plate at $M_\delta = 0$ (Blasius profile); see also Fig. 2.2, curve for $m = 0$, and eq. (2.43) for η.

to E. Pohlhausen [92] for a laminar boundary layer:

$$\frac{\delta_S}{\delta_T} \approx \sqrt{Pr} \approx 0.85 \text{ for air.} \tag{2.162}$$

2.3.8 Exact relations for heat transfer in flows without pressure and temperature gradients in the direction of flow

Van Driest's theory [which is strictly valid only under the assumptions $dp/dx = 0$, $\partial T/\partial x = 0$ $(dT_w/dx = 0)$] allows to compute directly the exchange of heat energy between the wall and the boundary layer. At a point x the heat energy transferred by conduction* per unit of time, unit

*At very high temperatures we have to consider heat transfer by radiation, which follows well-known laws (Stefan-Boltzmann equation); see the specific literature, as for example Eckert [19].

of length, and unit of width is given by

$$q_w(x) = -\left(\lambda \frac{\partial T}{\partial y}\right)_{y=0}.$$ (2.163)

For turbulent boundary layer $\lambda_{y=0}$ is also identical with the molecular heat conductivity because of $(\lambda_s)_{y=0} = 0$ [cf. eq. (1.24)].

The heat energy transferred along the length L per unit of time and per unit of width is then

$$Q = \int_0^L q_w(x)\,\mathrm{d}x.$$ (2.164)

The temperature gradient $(\partial T/\partial y)_{y=0}$ follows from eq. (2.137) if one recognizes that the assumptions in van Driest's theory require the temperature T to be a function of the velocity u only. This function depends on the Mach number M_δ, the heat transfer parameter b according to eq. (2.158), and the Prandtl number Pr. Equation (2.163) may therefore be written as follows (the subscript w indicating the point $y = 0$):

$$q_w(x) = -\left(\lambda \frac{\partial T}{\partial y}\right)_w = -c_p \left(\frac{\lambda}{\mu c_p}\right)_w \cdot \left(\frac{\mathrm{d}T}{\mathrm{d}u}\right)_w \cdot \left(\mu \frac{\partial u}{\partial y}\right)_w.$$ (2.165)

Assuming the Prandtl number to be constant, we have

$$\left(\frac{\lambda}{\mu c_p}\right)_w = \frac{1}{(\mathrm{Pr}_{Jw}} = \frac{1}{\mathrm{Pr}} = \text{const.}$$ (2.166)

(Pr = molecular Prandtl number, also for turbulent boundary layer). In eq. (2.165) the shear stress

$$\tau_w(x) = \left(\mu \frac{\partial u}{\partial y}\right)_w$$ (2.167)

occurs. The heat energy $q_w(x)$ is therefore proportional to the shear stress $\tau_w(x)$ which is known from the computation of the flow boundary layer. This result is called the *Reynolds analogy* between friction and heat transfer at the wall.

For computing $q_w(x)$, the derivative $(\mathrm{d}T/\mathrm{d}u)_w$ has to be obtained from eq. (2.137). We have

$$\left(\frac{\mathrm{d}T}{\mathrm{d}u}\right)_w = \frac{T_\delta}{u_\delta}\left[\frac{\mathrm{d}(T/T_\delta)}{\mathrm{d}(u/u_\delta)}\right]_w = \frac{T_\delta}{u_\delta}\frac{T_e - T_w}{T_\delta}\left[\frac{\mathrm{d}f_1}{\mathrm{d}(u/u_\delta)}\right]_w$$

$$= \frac{T_\delta}{u_\delta}\frac{T_e - T_w}{T_\delta} \cdot \frac{(\mathrm{Pr})_w}{s}.$$ (2.168)

Using eqs. (2.165) through (2.168), we find $q_w(x)$ in the form

$$q_w(x) = -\frac{1}{s} \cdot \frac{\tau_w}{\rho_\delta u_\delta^2} \frac{T_e - T_w}{T_\delta} \cdot \frac{c_p T_\delta}{u_\delta^2} \rho_\delta u_\delta^3 \qquad (2.169)$$

or in the dimensionless form

$$\frac{q_w(x)}{\rho_\delta u_\delta^3} = -\frac{c_f}{2s} \cdot \frac{T_e - T_w}{T_\delta} \cdot \frac{c_p T_\delta}{u_\delta^2}, \qquad (2.170)$$

where

$$c_f = \frac{2\tau_w}{\rho_\delta u_\delta^2}. \qquad (2.171)$$

The reason for calling the quantity s the Reynolds analogy factor is apparent from eq. (2.170): for Prandtl numbers other than unity, s expresses the influence of the Prandtl number on the connection between heat transfer and wall friction — a relation which is named after Osborne Reynolds. The ratio

$$\frac{c_f}{2s} = ST = \frac{q_w(x)}{\rho_\delta u_\delta c_p (T_e - T_w)} \qquad (2.172)$$

is also called the Stanton number. Many papers on problems of heat transfer used the Nusselt number Nu. Within the range of validity of the theory described here, that is for $dT_w/dx = 0$ and $dp/dx = 0$, the Nusselt number and the Stanton number are related to one another as follows:

$$Nu_x = \frac{q_w(x)x}{\lambda_w(T_e - T_w)} = ST \cdot Pr \cdot R_x \qquad (2.173)$$

where

$$R_x = \frac{\rho_\delta u_\delta x}{\mu_\delta}. \qquad (2.174)$$

For ideal gases (and air), eq. (2.170) may be further simplified by using the relation (2.135) from gas dynamics, and the parameter Θ [eq. (2.160)] for heat transfer. One obtains

$$\frac{q_w(x)}{\rho_\delta u_\delta^3} = -\frac{r}{2} ST \, \Theta. \qquad (2.175)$$

The problem of heat transfer is thus solved, whenever the local coefficient of friction c_f, and thus the Stanton number ST, is given along with the quantities r, s, and Θ, which are generally assumed to be known [cf. eqs. (2.150) through (2.153)].

In turbulent flow boundary layers, where the molecular Prandtl number Pr may differ strongly from unity in the laminar sublayer, von Kármán [55] has shown that

$$ST = \frac{c_f}{2}\left\{1 + 5\sqrt{\frac{c_f}{2}}[Pr - 1 + \ln(1 + \tfrac{5}{6}(Pr - 1))]\right\}^{-1}. \quad (2.176)$$

The influence of the molecular Prandtl number Pr on ST, for small values of c_f (i.e., large values of R_x), is according to this formula relatively minor and for gases with $0.7 < Pr < 1.2$ altogether negligible.

Based on the approximation theory which will be developed in the following section, we shall later find a relatively simple way to improve the computation of heat transfer in those cases when the assumptions of van Driest's theory, $dp/dx = 0$, $T_w = $ const. $(dT_w/dx = 0)$ are not satisfied. We shall deal with this so-called "modified Reynolds analogy" in Section 5.3.

3 Approximate Solutions in Boundary Layer Theory

3.1 Basic Concepts

Exact solutions for the problem of laminar boundary layers have thrown light on some basic properties of laminar velocity profiles in special cases, most of all in the study of similar solutions in accelerated and retarded flows of the type $u_\delta(x) \sim x^m$. Moreover, the development of special exact solutions for the simultaneous system of the differential equations for the flow and temperature boundary layers led to the recognition of certain coupling laws between the velocity and temperature profiles. It thus appears possible to develop general trial solutions for the velocity profile (with a known coupling relation to the temperature profile) for arbitrary free-stream flows in either a compressible or incompressible medium without or with heat transfer at the wall; these are expressions that possess at least the properties of the special solutions. It is then only necessary to provide for sufficient flexibility to allow adaptation of these expressions to fit general problems.

A first step in developing such solutions might consist in the introduction of a polynomial in y/δ for the velocity profile u/u_δ whose coefficients may be determined, for example, from the tables of similar solutions given by Hartree [43] (see Table 1 and Section 2.2.1.2). The exponent m of eq. (2.27) is then the only shape parameter of the velocity profile. Thus, all other coefficients of this polynomial solution are functions of m only. The class of special solutions of eq. (2.46) by Hartree can then be written in the form

$$\frac{u(x, y)}{u_\delta(x)} = f\left[\frac{y}{\delta(x)}, m\right]. \tag{3.1}$$

In order to make (3.1) suitable as a trial solution for more general problems, as for example of the type

$$u_\delta(x) \sim x^{m(x)}, \tag{3.2}$$

Table 3.1. *Hartree profiles. Solutions of the equation*
$$f''' + ff'' + \beta^*(1 - f'^2) = 0$$

[For definitions of α, Γ, and H, see eqs. (3.94), (3.96), and (3.75)]

β^*	−0.1988	−0.19	−0.18	−0.16	−0.14	−0.10	0
m	−0.0904	−0.0867	−0.0826	−0.0741	−0.0654	−0.0476	0.0000
α	0	0.0499	0.0732	0.1054	0.1293	0.1643	0.2212
Γ	−0.0683	−0.0639	−0.0581	−0.0490	−0.0408	−0.0265	0.0000
H	1.5150	1.5200	1.5250	1.5333	1.5400	1.5517	1.5720
η				$f' = u/u_\delta$			
0	0	0	0	0	0	0	0
0.1	0.0010	0.0095	0.0137_5	0.0198_5	0.0246_5	0.0324	0.0469_5
0.2	0.0040	0.0209	0.0293	0.0413	0.0507	0.0659	0.0939
0.3	0.0089	0.0343	0.0467	0.0643	0.0781	0.1003	0.1408
0.4	0.0158	0.0495	0.0659	0.0889	0.1069	0.1356	0.1876
0.5	0.0248	0.0665	0.0868	0.1151	0.1370	0.1718	0.2342
0.6	0.0358	0.0855	0.1094	0.1427	0.1684	0.2088	0.2806
0.7	0.0487	0.1063	0.1338	0.1719	0.2010	0.2466	0.3266
0.8	0.0636	0.1289	0.1598	0.2023	0.2347	0.2849	0.3720
0.9	0.0803	0.1533	0.1874	0.2341	0.2694	0.3237	0.4167
1.0	0.0991	0.1794	0.2166	0.2671	0.3050	0.3628	0.4606
1.2	0.1423	0.2364	0.2791	0.3362	0.3784	0.4415	0.5453
1.4	0.1927	0.2991	0.3463	0.4083	0.4534	0.5194	0.6244
1.6	0.2498	0.3665	0.4170	0.4820	0.5284	0.5948	0.6967
1.8	0.3126	0.4372	0.4896	0.5555	0.6016	0.6660	0.7610
2.0	0.3802	0.5095	0.5621	0.6269	0.6712	0.7314	0.8167
2.2	0.4509	0.5814	0.6327	0.6944	0.7354	0.7896	0.8633
2.4	0.5230	0.6509	0.6995	0.7561	0.7927	0.8398	0.9011
2.6	0.5946	0.7162	0.7605	0.8107	0.8422	0.8817	0.9306
2.8	0.6635	0.7754	0.8146	0.8574	0.8836	0.9153	0.9529
3.0	0.7278	0.8273	0.8607	0.8959	0.9168	0.9413	0.9691
3.2	0.7858	0.8713	0.8986	0.9265	0.9425	0.9607	0.9804
3.4	0.8364	0.9071	0.9286	0.9499	0.9616	0.9746	0.9880
3.6	0.8789	0.9352	0.9515	0.9669	0.9752	0.9841	0.9929
3.8	0.9132	0.9563	0.9681	0.9789	0.9845	0.9904	0.9959
4.0	0.9399	0.9716	0.9798	0.9871	0.9907	0.9944	0.9978
4.2	0.9598	0.9822	0.9876	0.9924	0.9946	0.9969	0.9988
4.4	0.9741	0.9893	0.9927	0.9957	0.9970	0.9983	0.9994
4.6	0.9839	0.9938	0.9959	0.9977	0.9984	0.9991	0.9997
4.8	0.9904	0.9965	0.9978	0.9988	0.9992	0.9996	0.9999
5.0	0.9945	0.9981_5	0.9988_5	0.9994	0.9996	0.9998	0.9999_5
5.2	0.9969	0.9990	0.9994	0.9997	0.9998	0.9999	1.0000
5.4	0.9984	0.9995	0.9997	0.9999	0.9999_5	1.0000	—
5.6	0.9992	0.9997	0.9999	0.9999_5	1.0000	—	—
5.8	0.9996_5	0.9999	0.9999_5	1.0000	—	—	—
6.0	0.9998_5	0.9999_5	1.0000	—	—	—	—
6.2	0.9999_5	1.0000	—	—	—	—	—
6.4	1.0000	—	—	—	—	—	—

one may consider the parameter m to be a function of x; eq. (3.1) is then written as

$$\frac{u(x, y)}{u_\delta(x)} = f\left[\frac{y}{\delta(x)}, m(x)\right].$$
(3.3)

Table 3.1. *Hartree profiles. Solutions of the equation*
$$f''' + ff'' + \beta^*(1 - f'^2) = 0$$

β^*	0.1	0.2	0.3	0.4	0.5	0.6	0.8
m	0.0526	0.1111	0.1765	0.2500	0.3333	0.4286	0.6666
α	0.2542	0.2775	0.2964	0.3132	0.3247	0.3337	0.3489
Γ	0.0187	0.0326	0.0439	0.0538	0.0612	0.0673	0.0776
H	1.5857	1.5950	1.6018	1.6070	1.6113	1.6150	1.6207

η				$f' = u/u_\delta$			
0.0	0	0	0	0	0	0	0
0.1	0.0582	0.0677	0.0760	0.0834	0.0903	0.0966	0.1080
0.2	0.1154	0.1334	0.1490	0.1628	0.1756	0.1872	0.2081
0.3	0.1715	0.1970	0.2189	0.2382	0.2558	0.2719	0.3003
0.4	0.2265	0.2584	0.2858	0.3097	0.3311	0.3506	0.3848
0.5	0.2803	0.3177	0.3495	0.3771	0.4015	0.4235	0.4619
0.6	0.3328	0.3747	0.4100	0.4403	0.4670	0.4907	0.5317
0.7	0.3839	0.4294	0.4672	0.4994	0.5276	0.5524	0.5947
0.8	0.4335	0.4816	0.5212	0.5545	0.5834	0.6086	0.6512
0.9	0.4815	0.5312	0.5718	0.6055	0.6344	0.6596	0.7015
1.0	0.5274	0.5782	0.6190	0.6526	0.6811	0.7056	0.7460
1.2	0.6135	0.6640	0.7033	0.7351	0.7615	0.7837	0.8194
1.4	0.6907	0.7383	0.7743	0.8027	0.8258	0.8449	0.8748
1.6	0.7583	0.8011	0.8326	0.8568	0.8760	0.8917	0.9154
1.8	0.8160	0.8528	0.8791	0.8988	0.9141	0.9264	0.9443
2.0	0.8637	0.8940	0.9151	0.9305	0.9421	0.9514	0.9644
2.2	0.9019	0.9260	0.9421	0.9537	0.9621	0.9689	0.9779
2.4	0.9315	0.9500	0.9617	0.9700	0.9760	0.9807	0.9867
2.6	0.9537	0.9672	0.9754	0.9812	0.9852	0.9884	0.9922
2.8	0.9697	0.9792	0.9847	0.9886	0.9913	0.9933	0.9956
3.0	0.9808	0.9873	0.9908	0.9933	0.9952	0.9962	0.9976
3.2	0.9883	0.9924	0.9946	0.9962	0.9974	0.9979	0.9987
3.4	0.9931	0.9957	0.9970	0.9979	0.9986	0.9989	0.9993
3.6	0.9961	0.9976	0.9984	0.9989	0.9993	0.9995	0.9997
3.8	0.9978	0.9987	0.9991_5	0.9994	0.9997	0.9997_5	0.9998_5
4.0	0.9988_5	0.9993	0.9995_5	0.9997	0.9999	0.9999	0.9999_5
4.2	0.9994	0.9996_5	0.9997_5	0.9999	0.9999_5	0.9999_5	1.0000
4.4	0.9997	0.9998_5	0.9999	0.9999_5	1.0000	1.0000	—
4.6	0.9998_5	0.9999_5	0.9999_5	1.0000	—	—	—
4.8	0.9999_5	1.0000	1.0000	—	—	—	—
5.0	1.0000	—	—	—	—	—	—

Equation (3.3) expresses the fact that for an arbitrary function $u_{\ddot{o}}(x)$ one considers those and only those velocity profiles which are contained in the "catalog" compiled by Hartree. Moreover, eq. (3.3) contains the assumption that the velocity profiles of the Hartree class can be characterized by the parameter m and the finite boundary layer thickness δ. With Section 1.5 in mind it should be said, however, that δ plays only the role of an auxiliary quantity whose definition does not influence the final result of the boundary layer computation in any essential way.

Table 3.1. *Hartree profiles. Solutions of the equation*
$$f''' + ff'' + \beta^*(1 - f'^2) = 0$$

β^*	1.0	1.2	1.6	2.0	
m	1.0	1.5	4.0000	∞	
α	0.3605	0.3681	0.3833	0.3922	
Γ	0.0856	0.0911	0.1016	0.1082	
H	1.6250	1.6290	1.6345	1.6380	
η	$f' = u/u_\delta$				
0.0	0	0	0	0	
0.1	0.1183	0.1276	0.1441	0.1588	
0.2	0.2266	0.2433	0.2726	0.2980	
0.3	0.3252	0.3475	0.3859	0.4186	
0.4	0.4144	0.4405	0.4849	0.5219	
0.5	0.4946	0.5231	0.5708	0.6096	
0.6	0.5662	0.5959	0.6446	0.6834	
0.7	0.6298	0.6596	0.7076	0.7449	
0.8	0.6859	0.7150	0.7610	0.7958	
0.9	0.7350	0.7629	0.8058	0.8376	
1.0	0.7778	0.8037	0.8432	0.8717	
1.2	0.8467	0.8682	0.8997	0.9214	
1.4	0.8968	0.9137	0.9375	0.9530	
1.6	0.9324	0.9450	0.9620	0.9726	
1.8	0.9569	0.9658	0.9775	0.9845	
2.0	0.9732	0.9793	0.9871	0.9914	
2.2	0.9841	0.9879	0.9928	0.9954	
2.4	0.9905	0.9931	0.9961	0.9976	
2.6	0.9946	0.9962	0.9980	0.9989	
2.8	0.9971	0.9980	0.9990	0.9994_5	
3.0	0.9985	0.9989	0.9995	0.9997_5	
3.2	0.9992	0.9995	0.9998	0.9999	
3.4	0.9996	0.9997_5	0.9999	1.0000	
3.6	0.9998	0.9999	1.0000	—	
3.8	0.9999	1.0000	—	—	
4.0	1.0000	—	—	—	

The solution of the problem of the *laminar* boundary layer with constant material properties, based on the trial solution (3.3), is obviously reduced to the simpler problem of determining the quantities $\delta(x)$ and $m(x)$ for a given free-stream flow $u_\delta(x)$ as a function of x.

For the determination of the two boundary layer parameters $\delta(x)$ and $m(x)$ we basically have the system of eqs. (2.20) and (2.21) available. But once the function f in eq. (3.3) is known, the dependence of the velocity u on the distance from the wall y is also given through the parameters $\delta(x)$ and $m(x)$, and a partial differential equation is no longer required to solve the problem thus simplified by eq. (3.3).

In connection with an approximate solution of a boundary layer problem of this kind,* von Kármán [56] and Pohlhausen [91] in 1921 for the first time made the fundamental proposal to average the system of eqs. (2.20) and (2.21) over the entire boundary layer thickness, that is, to integrate partially from $y = 0$ to $y = \delta$. The result is a so-called integral condition for the equilibrium of forces that act on a control section of width dx in the direction of flow and the height $y = \delta$ (see Fig. 3.1, p. 00). This momentum "integral condition" is an ordinary differential equation of first order in terms of quantities which result from the given function f by integration or differentiation with respect to y.

Since the trial solution (3.3) for f contains two unknowns, $\delta(x)$ and $m(x)$, it is obvious that the momentum integral condition does not suffice for the approximate solution of the boundary layer problem. As a second equation, Pohlhausen [91] used the condition, which follows from eq. (2.20) that at the wall, $y = 0$, only the pressure and viscous forces interact with one another, since the inertial forces vanish, because u equals zero there. This condition, also called "compatibility condition," is not a differential equation, but establishes (for incompressible flow) the connection between the second derivative $(\delta^2 u/\partial y^2)_{y=0}$ and the pressure gradient dp/dx directly.

This important statement about the shape (the curvature) of the velocity profile in the vicinity of the wall, as expressed by this compatibility condition, together with the momentum integral condition, affords a kind of selection principle according to which the velocity profiles offered by the trial solution (3.3) can be correlated with the length variable x (just as the letters of a typesetting machine on their return to the magazines).

---

*Since similar solutions of the system (2.20), (2.21) were not known at the time, the function f of the trial solution (3.3) for the velocity profile had to be roughly estimated, so that at least Blasius' velocity profile was approximated and physically plausible conditions were satisfied at the points $y = 0$ and $y = \delta$. The result was a polynomial of fourth degree in y/δ which served for decades as a rather useful starting point for computations of laminar incompressible boundary layers and which is still used today because of its mathematical simplicity.

Although this compatibility condition is a direct consequence of Prandtl's boundary layer equation (2.20), its value is obviously over-estimated if one expects the shape of the velocity profile in the immediate vicinity of the wall to determine its shape in the interior of the boundary layer. This fact reflects the practical limits of Pohlhausen's approximation theory, although its results are quite useful so long as the pressure gradient $\mathrm{d}p/\mathrm{d}x = -\rho_\delta u_\delta \, \mathrm{d}u_\delta/\mathrm{d}x$ changes little with x, or so long as the external velocity $u_\delta(x)$ is close to the law $u_\delta(x) \sim x^m$ of the similar solutions. For $u_\delta(x) \sim x^m$, the solution given by the approximation theory based on the Hartree velocity profiles in eq. (3.3) is even identical with the exact solution, as will be shown later (Section 6.3.2). When the free-stream velocity $u_\delta(x)$ varies strongly or periodically with x, this method may fail; this takes place more readily when the pressure increases than when it decreases in the direction of flow (see also the theories derived by Nickel [80]).

In order to improve the accuracy of this approximation theory with only one shape parameter for the velocity profile in eq. (3.3), it was suggested to develop an ordinary differential equation that is more powerful than the compatibility condition and can be used together with the momentum integral condition as a second equation.

Besides the conservation law for momentum, the conservation law for energy plays a major role in physics. It was therefore a logical step to improve the physical foundations of the approximation theory by adding the integral condition for the energy in the boundary layer. In this con-nection, a 1935 paper by Leibenson [65] should be mentioned, in which, probably for the first time, such an integral condition for energy was derived. Independent of that work, Wieghardt [155] in 1944 interpreted that integral condition and the already known integral condition for momentum as two of the infinitely many ordinary differential equations that can be derived from the system (2.20), (2.21) according to a general mathematical principle.

Wieghardt combined this with another important step to improve the approximation theory for the boundary layer by using — in addition to the integral conditions for momentum and energy — the compatibility condition and thereby establishing a simultaneous system of three equations. It was thus made possible to start out from a more general trial solution than (3.3), viz., a function with the two shape parameters $m_1(x)$ and $m_2(x)$:

$$\frac{u(x, y)}{u_\delta(x)} = F\left[\frac{y}{\delta(x)}, m_1(x), m_2(x)\right].$$ (3.4)

When this method was applied in practice, there were difficulties owing to a peculiarity of the polynomials of eleventh degree in y/δ. These difficulties

appeared when the pressure decreased in the direction of flow (mani-
fested by the occurrence of velocity profiles with values of $u/u_\delta > 1$,
which is physically impossible in the problems chosen by Wieghardt).*
However, this approximation theory produced remarkable improvements
over that of Pohlhausen for increasing pressure in the direction of flow,
particularly in the computation of the laminar separation point.

The difficulties in Wieghardt's theory have been explained and overcome by Geropp [33].† There is thus available for the laminar boundary
layer an approximation theory with two shape parameters for the velocity
profile and one parameter for the boundary layer thickness; this theory
produces remarkable agreement with the exact solution of the boundary
layer value problem when the integral conditions for momentum and
energy as well as the compatibility condition are simultaneously satisfied
(as borne out by experience gained so far with complicated problems).
Geropp was, moreover, able to generalize this theory to compressible flows
with heat transfer.

The accuracy of such an approximation theory can obviously be improved to any desired degree by imposing more integral conditions on the
system of equations than can, according to Wieghardt [155], be derived
by partially integrating the system (2.20), (2.21). For each additional
equation another shape parameter in the expression for the velocity
profile can be added, such that, for example, eq. (3.4) may be written with
j shape parameters:

$$\frac{u(x, y)}{u_\delta(x)} = F\left[\frac{y}{\delta(x)}, m_1(x), m_2(x), \ldots m_j(x)\right]. \qquad (3.5)$$

This enables one to approximate the exact solution to any desired accuracy.

For the solution of the boundary layer problem, based on a simultaneous
system of integral conditions, as sketched above, the question whether
the method converges as the number j of integral conditions increases is
of course of prime importance. It is not clear from the mathematical point
of view whether, for example, the integral conditions of momentum and
energy, among the infinitely many available integral conditions, are
actually most important (as one would expect from a physical point of
view). The question of convergence for this principle of solution, and the
relative importance of the different integral conditions is also discussed
in Geropp's paper [33]. He has shown the particular importance of the
integral conditions for momentum and energy (in the following referred
to as "momentum law" and "energy law," respectively). An investigation

*Cf. Nickel [82].
†Another proposal to improve on Wieghardt's theory was made by Head [44]. However, Geropp's paper is more general.

by Dorodnitsyn [14] deals with the reliability of the method of integral conditions.

The result in Geropp's paper makes it appear plausible that the approximation theory proposed by Walz [140] in 1944, which uses only the momentum and energy laws and a one-parameter trial solution (3.3), is generally more reliable than that of Pohlhausen [91]. Momentum law and compatibility condition are therefore, in general, of lower information content than the simultaneous system of the momentum and energy laws. This approximation theory with velocity profiles by Hartree [43], or others of type (3.3) which are of more or less equal value, has found wide applications in the last two decades. The Pohlhausen theory was used simultaneously because of its simplicity, usually in the form given by Holstein and Bohlen [47] or Koschmieder and Walz [59], who reduced it to a simple quadrature.

The principal ideas discussed in this section about approximate solutions of the flow-boundary layer problem and of the temperature boundary layer, the latter known by the coupling law (2.156), strictly apply to *laminar* boundary layers only. The train of thought developed here is, however, applicable to a large extent for *turbulent* boundary layers as well. At the present time it actually represents the only possible way to compute turbulent boundary layers for arbitrary external flows. The trial solutions (3.3) or (3.4) for the velocity profile (time averages for the velocity must be used here) have to be determined experimentally in this case. The integral conditions for momentum and energy are valid, subject only to minor restrictions (see, for example, Rotta [108]), just as in the case of laminar boundary layers. These equations contain, however, aside from the integrals in eqs. (3.3) or (3.4), also expressions for the shear stress at the wall and the work of the turbulent shear stresses (the dissipation integral), which must also be determined from experiment rather than from molecular viscosity (see also Section 1.3 on the "apparent" viscosity of turbulent boundary layers).

In compressible flow without and with heat transfer the determination of these laws poses special problems which will later be treated extensively (Sections 3.6 and 3.7) because of their practical importance.

The "compatibility condition," by the way, is also valid for turbulent boundary layers. Its influence on the turbulent velocity profile (on the shape parameter) however, is weaker than that for the laminar boundary layer. The influence of the compatibility condition on the computation of turbulent boundary layers is therefore as a rule comparatively small.

In this survey on the principal ideas of solving the boundary layer problem by approximation methods, we should mention that integral conditions may also be derived from the partial differential eqs. (2.89), (2.91) for the temperature (enthalpy) boundary layer, or from eq. (2.87) for the total energy. We shall use such an integral condition, for example, for purposes of generalizing the coupling law (2.156), in order to derive the "modified" Reynolds analogy (Section 5.3). It should also be mentioned here, that it is possible to derive infinitely many "compatibility conditions" from eqs. (2.20), (2.89), or (2.91) by partial differentiation. In analogy to the case of systems of integral conditions, the compatibility conditions of "lower order" are more important than those of higher order. Section 3.4 deals more explicitly with some of the compatibility conditions that are useful for approximation theories.

3.2 Integral Conditions Derived from Prandtl's Boundary Layer Equation

3.2.1 The integral condition for momentum (physical derivation)

Prandtl's boundary layer eq. (2.20) for two-dimensional flow was derived on the basis of the boundary layer simplifications (Section 1.7) by formulating the balance of forces that act on an infinitesimal control section $dx\, dy$ of unit width.

The integral condition for momentum results, aside from the purely mathematical principle of partial integration which will be explained later, in Fig. 3.1 by writing down the balance of forces for a control section $dx \cdot \delta \cdot 1$ which extends in y direction to the distance $y = \delta$. Owing to Prandtl's boundary layer simplifications, only forces in x direction are to be considered.

We begin by formulating the *continuity condition*. Through 1, 2 (Fig. 3.1) the following mass flows into the control section per unit of time and per unit of width:

$$\int_0^\delta \rho u \, dy. \tag{3.6}$$

Through 3, 4 the mass

$$-\left[\int_0^\delta \rho u \, dy + \frac{d}{dx} \left(\int_0^\delta \rho u \, dy \right) dx \right] \tag{3.7}$$

flows out of the control section. The change of mass flow is therefore given by

$$-\frac{d}{dx} \left(\int_0^\delta \rho u \, dy \right) dx. \tag{3.8}$$

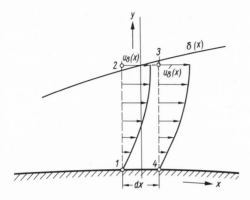

Fig. 3.1. Illustration of the partial integration of Prandtl's boundary layer eq. (2.20) with the aid of the continuity eq. (2.21).

This change of mass flow has to be equalized, for reasons of continuity, by an equivalent mass flow in y direction with the velocity v_δ through the side 2, 3. Thus, the continuity condition for the control section 1, 2, 3, 4 becomes

$$\frac{\mathrm{d}}{\mathrm{d}x}\left(\int_0^\delta \rho u\,\mathrm{d}y\right)\mathrm{d}x = -\rho_\delta v_\delta\,\mathrm{d}x. \tag{3.9}$$

The *equilibrium* of *forces* in turn is established by the participation (excluding buoyant forces) of *inertial forces* (through changes of momentum), *pressure forces*, and *frictional forces*.

The following *momentum* enters through 1, 2 in x direction per unit of time and per unit of width

$$\int_0^\delta \rho u^2\,\mathrm{d}y \tag{3.10}$$

(positive sign); the momentum

$$-\left[\int_0^\delta \rho u^2\,\mathrm{d}y + \frac{\mathrm{d}}{\mathrm{d}x}\left(\int_0^\delta \rho u^2\,\mathrm{d}y\right)\mathrm{d}x\right] \tag{3.11}$$

leaves through 3, 4 (negative sign). Thus, there is a net change of momentum

$$-\frac{\mathrm{d}}{\mathrm{d}x}\left(\int_0^\delta \rho u^2\,\mathrm{d}y\right)\mathrm{d}x. \tag{3.12}$$

This, however, is not yet the total change of momentum in x direction in the control section. The mass flow $\rho_\delta v_\delta \, dx$, given by eq. (3.9), carries with it the momentum

$$-\rho_\delta v_\delta u_\delta \, dx = u_\delta \frac{d}{dx}\left(\int_0^\delta \rho u \, dy\right) dx \tag{3.13}$$

through 2, 3 out of the control section. The total change of momentum in x direction is therefore given by

$$\left[-\frac{d}{dx}\left(\int_0^\delta \rho u^2 \, dy\right) + u_\delta \frac{d}{dx}\left(\int_0^\delta \rho u \, dy\right)\right] dx. \tag{3.14}$$

The *pressure force* per unit width on the side 1, 2 of the control section is given by

$$\delta \cdot p. \tag{3.15}$$

The pressure

$$-\delta\left(p + \frac{dp}{dx}\,dx\right). \tag{3.16}$$

acts on the side 3, 4. The resulting pressure force per unit width on the control section is therefore, after application of eq. (2.26),

$$-\delta\frac{dp}{dx}\,dx = \delta\rho_\delta u_\delta \frac{du_\delta}{dx}\,dx. \tag{3.17}$$

Finally, we have to formulate the *viscous force* that acts on the control section in x direction. At the upper edge 2, 3, because of $(\partial u/\partial y)_{y=0} = 0$, the shear stress $\tau_\delta = (\mu\,\partial u/\partial y)_{y=\delta} = 0$. Therefore, only the shear stress at the wall $\tau_w = (\mu\,\partial u/\partial y)_{y=0}$ acts along the line 1, 4. This results in a (negative) retarding force per unit width

$$-\tau_w \, dx. \tag{3.18}$$

The sum of the three forces (3.14), (3.17), and (3.18) must therefore be zero. Thus, we have the "momentum integral condition"

$$u_\delta \frac{d}{dx}\left(\int_0^\delta \rho u \, dy\right) - \frac{d}{dx}\left(\int_0^\delta \rho u^2 \, dy\right) + \delta\rho_\delta u_\delta \frac{du_\delta}{dx} - \tau_w = 0. \tag{3.19}$$

This relation can be translated into a form more useful for practical computation by observing that the term with the factor δ may, purely formally, be rewritten as follows:

$$\delta\rho_\delta u_\delta \frac{du_\delta}{dx} = -u_\delta \frac{d}{dx}(\rho_\delta u_\delta \delta) + \frac{d}{dx}(\rho_\delta u_\delta^2 \delta)$$

$$= -u_\delta \frac{d}{dx}\left(\rho_\delta u_\delta \int_0^\delta dy\right) + \frac{d}{dx}\left(\rho_\delta u_\delta^2 \int_0^\delta dy\right). \tag{3.20}$$

Substituting eq. (3.20) into (3.19) results first in

$$-u_\delta \frac{d}{dx}\left[\rho_\delta u_\delta \int_0^\delta \left(1 - \frac{\rho u}{\rho_\delta u_\delta}\right)dy\right]$$

$$+ \frac{d}{dx}\left[\rho_\delta u_\delta^2 \int_0^\delta \left(1 - \frac{\rho}{\rho_\delta}\left(\frac{u}{u_\delta}\right)^2\right)dy\right] - \tau_w = 0. \tag{3.21}$$

The second term in (3.21) can be rewritten by carrying out the differentiations

$$\frac{d}{dx}\left\{\rho_\delta u_\delta^2 \int_0^\delta \left[1 - \frac{\rho}{\rho_\delta}\left(\frac{u}{u_\delta}\right)^2\right]dy\right\} = u_\delta \frac{d}{dx}\left\{\rho_\delta u_\delta \int_0^\delta \left[1 - \frac{\rho}{\rho_\delta}\left(\frac{u}{u_\delta}\right)^2\right]dy\right\}$$

$$+ \rho_\delta u_\delta \frac{du_\delta}{dx} \int_0^\delta \left[1 - \frac{\rho}{\rho_\delta}\left(\frac{u}{u_\delta}\right)^2\right]dy. \tag{3.22}$$

Using the definitions

$$\int_0^\delta \left(1 - \frac{\rho}{\rho_\delta}\frac{u}{u_\delta}\right)dy = \delta_1 = \text{Displacement thickness} \tag{3.23}$$

and

$$\int_0^\delta \frac{\rho}{\rho_\delta}\frac{u}{u_\delta}\left(1 - \frac{u}{u_\delta}\right)dy = \delta_2 = \text{Momentum-loss thickness}, \tag{3.24}$$

we have

$$\int_0^\delta \left[1 - \frac{\rho}{\rho_\delta}\left(\frac{u}{u_\delta}\right)^2\right]dy = \delta_1 + \delta_2, \tag{3.25}$$

and (3.21) becomes

$$u_\delta \frac{d}{dx}(\rho_\delta u_\delta \delta_2) + \rho_\delta u_\delta^2(\delta_1 + \delta_2)\frac{1}{u_\delta}\frac{du_\delta}{dx} - \tau_w = 0. \tag{3.26}$$

The first expression in (3.26) can be transformed further by differentiation:

$$u_\delta \frac{d}{dx}(\rho_\delta u_\delta \delta_2) = \rho_\delta u_\delta^2 \frac{d\delta_2}{dx} + \rho_\delta u_\delta^2 \delta_2 \frac{1}{\rho_\delta} \frac{d\rho_\delta}{dx} + \rho_\delta u_\delta^2 \delta_2 \frac{du_\delta/dx}{u_\delta}$$

$$= \rho_\delta u_\delta^2 \left[\frac{d\delta_2}{dx} + \delta_2 \frac{d\rho_\delta/dx}{\rho_\delta} + \delta_2 \frac{du_\delta/dx}{u_\delta}\right]. \tag{3.27}$$

Substituting eq. (3.27) into (3.26) and dividing the entire equation by $\rho_\delta u_\delta^2$ yields, in nondimensional notation (δ_1 and δ_2 are lengths), the following ordinary first-order differential equation for the momentum-loss thickness δ_2:

$$\frac{d\delta_2}{dx} + \delta_2 \frac{du_\delta/dx}{u_\delta}\left[2 + \frac{\delta_1}{\delta_2} - M_\delta^2\right] - \frac{\tau_w}{\rho_\delta u_\delta^2} = 0. \tag{3.28}$$

Here the well-known gas-dynamical relation

$$\frac{1}{\rho_\delta}\frac{d\rho_\delta}{dx} = -\frac{1}{u_\delta}\frac{du_\delta}{dx}M_\delta^2 \tag{3.29}$$

was used (see Appendix II).

The "momentum integral condition" (3.28) — hereafter called the *momentum law* — has the nice property that the physically not very well defined boundary layer thickness is altogether eliminated. The quantities δ_1 and δ_2 which occur in this equation are practically independent of the definition of δ, because the integrands in eqs. (3.23), (3.24) which define these quantities are practically zero for $y \geq \delta$.

It is very easy to formulate also an integral condition for the mechanical energies in the control section 1, 2, 3, 4 of Fig. 3.1. Integral conditions for the enthalpy i and the total energy, as well as integral conditions of "higher order," may be derived physically in the same way as in the case of the momentum law, which for didactical reasons was shown in great detail. Transformations of the kind used with eqs. (3.20), (3.22), and (3.27) play an essential role in this. This more intuitive way to derive integral conditions may be replaced by the purely mathematical technique of partial integration of the differential eqs. (2.20), (2.21), (2.89), or (2.87) from $y = 0$ to the limit $y = \delta$. This even results in systems of equations with infinitely many integral conditions. All these equations are ordinary differential equations for unknown boundary layer integral quantities, such as δ_1 and δ_2, whose determination, based on trial solutions of type (3.3) for the velocity profiles and on an expression of type (2.156) for the temperature profile, is possible with known methods of mathematics.

3.2.2 *Derivation of a system of infinitely many integral conditions from Prandtl's boundary layer equation*

Based on a mathematical principle that is actually well known and was already indicated by Pohlhausen [91] and explained extensively by Wieghardt [155] in connection with Prandtl's boundary layer equation, it is possible to derive from a partial differential equation an infinite set of ordinary differential equations by taking moments (i.e., multiplying by "weight functions" and subsequent partial integration). Among these equations, the momentum integral condition is only a special case.

According to Wieghardt [155], it is useful to choose as "weight functions" powers of the velocity.* These powers may be integers or fractional, positive or negative. The rule for obtaining the system of integral conditions from (2.20), (2.21) is then given as follows:

Equation (2.20) is multiplied by u^ν:

$$u^\nu \left\{ \rho u \frac{\partial u}{\partial x} + \rho v \frac{\partial u}{\partial y} \right\} = \frac{1}{\nu + 1} \left\{ \rho u \frac{\partial u^{\nu+1}}{\partial x} + \rho v \frac{\partial u^{\nu+1}}{\partial y} \right\}$$

$$= u^\nu \left\{ -\frac{dp}{dx} + \frac{\partial \tau}{\partial y} \right\}. \tag{3.30}$$

Equation (2.21) is multiplied by $u^{\nu+1}/(\nu + 1)$:

$$\frac{u^{\nu+1}}{\nu + 1} \left\{ \frac{\partial(\rho u)}{\partial x} + \frac{\partial(\rho v)}{\partial y} \right\} = 0. \tag{3.31}$$

Addition of eqs. (3.30) and (3.31) results in

$$\frac{1}{\nu + 1} \left\{ \frac{\partial(\rho u^{\nu+2})}{\partial x} + \frac{\partial(\rho v u^{\nu+1})}{\partial y} \right\} = -u^\nu \left\{ \frac{dp}{dx} - \frac{\partial \tau}{\partial y} \right\}. \tag{3.32}$$

This equation is now integrated partially with respect to y from $y = 0$ to $y = \delta$. Observing eq. (2.26), we obtain the relation

$$\frac{1}{\nu + 1} \left[\int_0^\delta \frac{\partial(\rho u^{\nu+2})}{\partial x} \, dy + \int_0^\delta \frac{\partial(\rho v u^{\nu+1})}{\partial y} \, dy \right]$$

$$= \rho_\delta u_\delta \frac{du_\delta}{dx} \int_0^\delta u^\nu \, dy + \int u^\nu \frac{\partial \tau}{\partial y} \, dy. \tag{3.33}$$

In order to include in this system of integral conditions also the limiting case $\nu = \infty$, it is necessary, as Geropp [36] pointed out, to retain the

*The powers of the distance from the wall, y, as weight functions lead also to meaningful integral conditions; cf. for example, Hudimoto [51].

differentials under the integrals on the left-hand side of eq. (3.33). Multiplication of eq. (3.33) by $(\nu + 1)/u_\delta^{\nu+2}$ renders that equation dimensionless. After that, some of the integrands have to be modified — in analogy to the derivation of the momentum integral condition (3.28) — such that these integrands tend to zero for $y \geq \delta$. Using the integrals

$$e_\nu = (\nu + 1) \int_0^\delta \left(\frac{u}{u_\delta}\right)^\nu \frac{\partial}{\partial y} \left(\frac{\tau}{\rho_\delta u_\delta^2}\right) dy, \tag{3.34}$$

$$f_\nu = \int_0^\delta \frac{\rho u}{\rho_\delta u_\delta} \left[1 - \left(\frac{u}{u_\delta}\right)^{\nu+1}\right] dy, \tag{3.35}$$

$$g_\nu = (\nu + 1) \int_0^\delta \frac{\rho u}{\rho_\delta u_\delta} \left[\frac{\rho_\delta}{\rho} \left(\frac{u}{u_\delta}\right)^{\nu-1} - 1\right] dy, \tag{3.36}$$

and the relation (3.29) from gas dynamics, one can write eq. (3.33) as follows:

$$\frac{df_\nu}{dx} + [(2 + \nu - M_\delta^2)f_\nu + g_\nu] \frac{1}{u_\delta} \frac{du_\delta}{dx} + e_\nu$$

$$+ \left\{\int_0^\delta \frac{\partial}{\partial x}\left[\frac{\rho u}{\rho_\delta u_\delta}\right] dy - \frac{1}{\rho_\delta u_\delta} \int_0^\delta \frac{\partial(\rho u)}{\partial x} dy - \int_0^\delta \frac{\partial}{\partial x}\left[\frac{\rho}{\rho_\delta}\left(\frac{u}{u_\delta}\right)^{\nu+2}\right] dy\right.$$

$$\left. - \int_0^\delta \frac{\partial}{\partial y}\left[\frac{\rho}{\rho_\delta} \frac{v}{u_\delta}\left(\frac{u}{u_\delta}\right)^{\nu+1}\right] dy - \frac{df_\nu}{dx}\right\} = 0. \tag{3.37}$$

Observing the rules for differentiating an integral with variable upper limit, it is possible to interchange differentiation and integration symbols of the first integral inside the braces. The second integral inside the braces can be evaluated with the aid of the continuity equation. The expressions inside the braces thus simplify as follows:

$$h_\nu = \frac{v_\delta}{u_\delta} - \frac{d\delta}{dx} + \frac{d}{dx} \int_0^\delta \frac{\rho}{\rho_\delta}\left(\frac{u}{u_\delta}\right)^{\nu+2} dy - \int_0^\delta \frac{\partial}{\partial x}\left[\frac{\rho}{\rho_\delta}\left(\frac{u}{u_\delta}\right)^{\nu+2}\right] dy$$

$$- \int_0^\delta \frac{\partial}{\partial y}\left[\frac{\rho v}{\rho_\delta u_\delta}\left(\frac{u}{u_\delta}\right)^{\nu+1}\right] dy. \tag{3.38}$$

The most general form for the system of integral conditions is then, after Geropp [36], given by

$$\frac{df_\nu}{dx} + f_\nu\left(2 + \nu - \frac{g_\nu}{f_\nu} - M_\delta^2\right) \frac{1}{u_\delta} \frac{du_\delta}{dx} + e_\nu + h_\nu = 0. \tag{3.39}$$

As long as $\nu \neq \infty$, the integrands of h_ν are different from zero for $y > 0$, because of $(u/u_\delta)^\nu > 0$. The interchange of differentiation and integration is permissible, and thus $h_\nu = 0$ for all $\nu \neq \infty$.

The system (3.39) assumes in this case the well-known form given by Truckenbrodt [138] and Walz [153].

Since ν may assume any arbitrary value, eq. (3.39) represents a system of infinitely many ordinary differential equations for the infinitely many unknown boundary layer quantities in a trial solution of the type (3.5), all of which occur implicitly in the integral expressions e_ν, f_ν, and g_ν. It is easily seen that eq. (3.39) with (3.36) through (3.38) has different information content for each value of the subscript ν. For $\nu = 0$, eq. (3.39) is identical with the momentum integral condition (3.28). It follows from eqs. (3.36) through (3.38) for $\nu = 0$ that

$$e_0 = -\frac{\tau_w}{\rho_\delta u_\delta^2} \equiv -c_f/2; \tag{3.40}$$

$$f_0 = \delta_2; \tag{3.41}$$

$$g_0 = \delta_1; \tag{3.42}$$

$$h_0 = 0. \tag{3.42a}$$

Since multiplication of eq. (2.20) for the balance of forces by the velocity u leads to the energy eq. (2.88), we obtain for $\nu = 1$ the *energy integral condition* (for the mechanical energies). In this case we have

$$e_1 = \frac{2}{\rho_\delta u_\delta^3} \int_0^\delta u \frac{\partial \tau}{\partial y} \, dy = \frac{2}{\rho_\delta u_\delta^3} \left([u\tau]_0^\delta - \int_0^{u_\delta} \tau \, du \right)$$

$$= -2 \frac{\displaystyle\int_0^{u_\delta} \tau \, du}{\rho_\delta u_\delta^3} \equiv c_D = \text{dissipation integral}, \tag{3.43}$$

$$f_1 = \int_0^\delta \frac{\rho u}{\rho_\delta u_\delta} \left[1 - \left(\frac{u}{u_\delta}\right)^2 \right] dy \equiv \delta_3 = \text{energy-loss thickness}, \tag{3.44}$$

$$\tfrac{1}{2} g_1 = \int_0^\delta \frac{\rho u}{\rho_\delta u_\delta} \left(\frac{\rho_\delta}{\rho} - 1\right) dy \equiv \delta_4 = \text{density-loss thickness}, \tag{3.45}$$

$$h_1 = 0, \tag{3.45a}$$

and eq. (3.39) assumes the form:

$$\frac{d\delta_3}{dx} + \delta_3 \frac{du_\delta/dx}{u_\delta} \left(3 + 2\frac{\delta_4}{\delta_3} - M_\delta^2\right) - \frac{2}{\rho_\delta u_\delta^3} \int_0^{u_\delta} \tau \, du = 0. \tag{3.46}$$

From the mathematical point of view, the subscript ν, as indicated before, may assume any arbitrary numerical value, even the value $\nu = \infty$. In this case, because $0 < u/u_\delta < 1$,

$$\left[\left(\frac{u}{u_\delta} \right)^{\nu+2} \right]_{\nu \to \infty} \to 0, \tag{3.47}$$

and thus (after elementary computation),

$$[f_\nu(\nu + 2 - M_\delta^2) + g_\nu]_{\nu \to \infty} \to f_\infty(1 - M_\delta^2); \tag{3.48}$$

$$f_\infty = \int_0^\delta \frac{\rho u}{\rho_\delta u_\delta} \, dy = \delta - \delta_1; \ [\text{see eq. (3.23)}] \tag{3.49}$$

$$e_\infty = \left[\frac{\nu + 1}{\rho_\delta u_\delta^2} \int_0^\delta \left(\frac{u}{u_\delta} \right)^\nu \frac{\partial \tau}{\partial y} \, dy \right]_{\nu \to \infty} = 0; \tag{3.50}$$

$$h_\infty = \frac{v_\delta}{u_\delta} - \frac{d\delta}{dx}; \tag{3.51}$$

and finally the ordinary differential equation

$$\frac{df_\infty}{dx} + f_\infty(1 - M_\delta^2) \frac{du_\delta/dx}{u_\delta} + \frac{v_\delta}{u_\delta} - \frac{d\delta}{dx} = 0. \tag{3.52}$$

In this "degenerate" boundary layer equation there is no longer a frictional term. Because of relation (3.29) we may rewrite eq. (3.52) as follows:

$$\frac{d}{dx} [\rho_\delta u_\delta(\delta - \delta_1)] + \rho_\delta v_\delta - \rho_\delta u_\delta \frac{d\delta}{dx} = 0. \tag{3.53}$$

This equation is nothing but a continuity condition, which may also be derived directly by partial integration of the continuity eq. (2.7) over the entire boundary layer from $y = 0$ to $y = \delta$.

The function f_∞/δ, i.e., the difference $1 - \delta_1/\delta$, was determined empirically by Head [44] as a function of the shape parameter H_{12} for turbulent boundary layers. He derived the entrainment equation empirically from eq. (3.52). That entrainment equation, together with the momentum law (3.28), yields remarkably good predictions about the shape parameter H_{12} [or the quantity $H(H_{12})$] in incompressible flows with strong pressure gradients; this is significant, for example, in determining the separation point. Recently, Felsch [28] was able to show that the simultaneous system of differential equations (3.28), (3.46) yields better results than the method based on the entrainment equation (see also Section 6.9).

It may thus be said that in eq. (3.39) the integral condition for $\nu = \infty$ degenerates into a continuity condition that is no longer capable of giving any information about boundary layer properties. One may therefore expect that the integral conditions of low order [for example, for $\nu = 0$ (momentum law) and $\nu = 1$ (energy law)] have particularly high information content about the boundary layer properties. A more rigorous argument for the importance of the equations of low order can be made after investigating the dependence of the weight function u^ν [or in dimensionless notation $(u/u_\delta)^\nu$] on the subscript ν (Fig. 3.2).

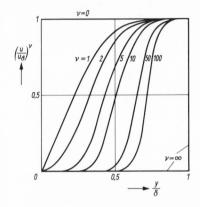

Fig. 3.2. The dependence of the weight function $(u/u_\delta)^\nu$ upon the order ν for the Blasius profile.

For $\nu = 0$, the weight function $(u/u_\delta)^\nu$ becomes $(u/u_\delta)^0 = 1$. After multiplication the individual terms of the boundary layer equation (2.20) retain their full information content. For $\nu = 1$, the weight function $(u/u_\delta)^\nu = (u/u_\delta)^1 = u/u_\delta$ at the wall ($y = 0$) is also zero. In the vicinity of the wall we have, already for the case $\nu = 1$, a certain reduction in the information content of eq. (3.39). With increasing values of ν, the weight function $(u/u_\delta)^\nu$ tends toward zero because of $0 < u/u_\delta < 1$ in a domain near the wall which increases in size. This means that the information content of the integral conditions is reduced with increasing values of ν. As $\nu \to \infty$, all characteristics of the boundary layer are suppressed in eq. (2.20) on the domain $0 < u/u_\delta < 1$ because of $(u/u_\delta)^\nu_{\nu \to \infty} = 0$, so that only the trivial continuity statement (3.53) remains.

Geropp [33] has investigated the dependence of the information content of the integral condition as a function of the order ν in an even more detailed way on the basis of special trial solutions for the velocity profile u/u_δ with the result, already mentioned earlier, that the integral conditions for the momentum [$\nu = 0$, eq. (3.28)] and for energy [$\nu = 1$, eq. (3.46)] have to be considered as particularly important equations of the system (3.39). These two equations are distinguished from the other equations of the system (3.39) also by the fact that they have physical meaning. It is therefore understandable that approximate solutions for the boundary layer problem have up to now been based exclusively on the application of these two equations (together with the "compatibility conditions").

From a mathematical point of view, however, other equations of the system (3.39) with low exponent v, such as $v = \frac{1}{4}, \frac{1}{2}, \frac{3}{4}$ should be considered no less useful. One should also expect that the approximate solution of the boundary layer problem turns out better as more equations of the system (3.39) are satisfied simultaneously. This poses the question, already brought up in Section 3.1, whether the approximate solution obtained from j equations (with a trial solution for the velocity profile containing $j + 1$ unknowns) converges to the exact solution of the system of partial differential eqs. (2.20), (2.21) as $j \to \infty$.

A mathematically satisfying answer to this question has yet to be found. Such convergent behavior should certainly be expected when the expression for the velocity profiles is determined in such a way that as many boundary conditions as possible are satisfied which derive from the system (2.20), (2.21) at the points $y = 0$ (compatibility conditions) and $y = \delta$ (or $y = \infty$) as exact consequences.

The formulation of such boundary conditions as additional equations which may play an important role in establishing approximation theories, along with integral conditions, will be the subject of Section 3.4. In Section 3.3 another system of infinitely many integral conditions will be derived which can be obtained from the partial differential eq. (2.87) for the total energy h.

3.3 Integral Conditions Derived from the Partial Differential Equation for the Total Energy

By multiplying eq. (2.87) by h^v, and the continuity eq. (2.21) by $h^{v+1}/(v + 1)$, and adding these equations one obtains

$$\frac{1}{v+1}\left\{\frac{\partial(\rho u h^{v+1})}{\partial x} + \frac{\partial(\rho v h^{v+1})}{\partial y}\right\} = h^v\left[\frac{\partial(u\tau)}{\partial y} + \frac{\partial}{\partial y}\left(\lambda\frac{\partial T}{\partial y}\right)\right], \quad (3.54)$$

where the number v is arbitrary. Partial integration with respect to y from $y = 0$ to $y = \delta$, in consideration of the dependence of δ on x, yields first

$$\frac{d}{dx}\left[\int_0^{\delta(x)} \rho u h^{v+1}\, dy\right] - \rho_\delta u_\delta h_\delta^{v+1}\cdot\frac{d\delta}{dx} + \rho_\delta v_\delta h_\delta^{v+1}$$

$$= (v + 1)\left\{\int_0^{\delta(x)} h^v\frac{\partial(u\tau)}{\partial y}\, dy + \int_0^{\delta(x)} h^v\frac{\partial}{\partial y}\left(\lambda\frac{\partial T}{\partial y}\right) dy\right\}. \quad (3.55)$$

The expression $\rho_\delta v_\delta$ may be replaced by using the continuity eq. (2.21). The terms with the factor $d\delta/dx$ cancel out. After elementary transformations aimed at eliminating the influence of the boundary layer thickness δ, as used in Sections 3.2.1 and 3.2.2, eq. (3.55) becomes

$$\frac{d}{dx}\left[\int_0^\delta \rho u(h^{\nu+1} - h_\delta^{\nu+1})\, dy\right]$$

$$= (\nu + 1)\left\{\int_0^\delta h^\nu \frac{\partial(u\tau)}{\partial y}\, dy + \int_0^\delta h^\nu \frac{\partial}{\partial y}\left(\lambda \frac{\partial T}{\partial y}\right)\, dy\right\}. \tag{3.56}$$

For $\nu = 0$ one obtains the integral condition for the total flux of energy:

$$\frac{d}{dx}\left[\int_0^\delta \rho u(h - h_\delta)\, dy\right] = [u\tau]_{y=0}^{y=\delta} - \left(\lambda \frac{\partial T}{\partial y}\right)_{y=0}$$

$$= -\left(\lambda \frac{\partial T}{\partial y}\right)_{y=0}. \tag{3.57}$$

This equation has the remarkable property that the term $[u\tau]_0^\delta$ which describes the work of the shear stresses, including dissipation, vanishes at $y = 0$ because of $u = 0$, and at $y = \delta$ because $\tau = 0$; this is even the case when the Prandtl number Pr is different from unity. This result differs from relation (2.102) in which that term vanishes only for Pr $= 1$.

In the system (3.56), which contains infinitely many integral conditions, only eq. (3.57), for $\nu = 0$, is physically interpretable. The other equations, however, are principally available for improvements of the approximation theory. We shall use eq. (3.57) in Section 5.3 to derive a generalization of eq. (2.156) for the purpose of improving the calculation of heat transfer. It should be noted here that for the case of incompressible flow, eq. (3.57) becomes the so-called "heat flux equation." For incompressible flow, the kinetic energy $u^2/2$ is negligible against the enthalpy i as can be seen immediately from eq. (2.113). Thus we have

$$(h)_{M_\delta \to 0} = c_p T, \tag{3.58}$$

and eq. (3.57) simplifies to

$$c_p \frac{d}{dx}\left[\int_0^\delta \rho u(T - T_\delta)\, dy\right] = -\left(\lambda \frac{\partial T}{\partial y}\right)_{y=0}. \tag{3.59}$$

Equation (3.59) expresses the fact that the enthalpy can be changed only by heat flow through the wall $y = 0$. At the outer edge of the boundary layer no heat can be exchanged with the free-stream flow (for example, with the flow in the core of pipes) since there $(\partial T/\partial y)_{y=\delta} = 0$. All heat entering the flow through the wall remains within the boundary layer (which grows with x).

It should be pointed out that a different system of integral conditions can be obtained from eq. (2.91) for the temperature field. In an earlier paper, Walz [141] used one equation of that system in connection with eq. (3.28) for approximate computations of compressible laminar and turbulent boundary layers. Experience showed, however, that computational practice is simplified when the system (3.28), (3.46) is used.

3.4 Boundary Conditions for Flow and Temperature Boundary Layers

3.4.1 General remarks

From the partial differential equations for the flow and temperature boundary layers there follow certain connections between the boundary layer quantities and the pressure or temperature field both at the wall $y = 0$ and at the outer edge $y = \delta$,* which are quite useful as additional (usually very simple) equations in the development of the approximation theories. The most important equations of this kind shall be presented here.

The boundary conditions at the point $y = \delta$ are trivial because of the asymptotic transition of the boundary layer velocity u toward u_δ and of the boundary layer temperature T toward T_δ. They are:

$$y = \delta: \; u = u_\delta; \; \frac{\partial u}{\partial y} = \frac{\partial^2 u}{\partial y^2} = \cdots = \frac{\partial^n u}{\partial y^n} = 0, \qquad (3.60)$$

$$y = \delta: \; T = T_\delta; \; \frac{\partial T}{\partial y} = \frac{\partial^2 T}{\partial y^2} = \cdots = \frac{\partial^n T}{\partial y^n} = 0. \qquad (3.61)$$

When the order n of the derivative of u in eq. (3.60) is even, the algebraic sign of the derivative before reaching the asymptote $u = u_\delta$ is negative; for odd n it is positive. No general statement can be made about the algebraic sign of the derivatives of T before reaching the asymptote $T = T_\delta$, however. The heat transfer at the wall will here, in general, have an unpredictable influence on the algebraic sign of $(\partial^n T/\partial y^n)_{y=\delta}$.

The predictive value of this condition at the outer edge of the boundary layer is generally quite small for purposes of an approximation theory. The boundary conditions for $y = 0$, the so-called "compatibility condi-

---

*As to the definition of the boundary layer thickness δ and the reasons for abandoning the boundary condition $y \to \infty$, cf. Section 1.5.

tions" have however — at least for laminar boundary layers — a relatively high predictive value for the behavior of $u(x, y)$ and $T(x, y)$ in the vicinity of the wall, comparable to that of the integral conditions from (3.39) and (3.56) for low orders ν.

3.4.2 Compatibility conditions for the flow boundary layer

At the wall $y = 0$, there is $u = 0$ and (as long as no flowing material is removed by suction or added by injection at the wall) also $v = 0$. Thus, Prandtl's boundary layer equation (2.20) reduces to the statement that only pressure and frictional forces interact at the wall:

$$\left[\frac{\partial}{\partial y}\left(\mu \frac{\partial u}{\partial y}\right)\right]_{y=0} = \frac{dp}{dx}. \tag{3.62}$$

After differentiating the left-hand side of eq. (3.62) and observing eq. (2.26), one obtains

$$\left[\mu \frac{\partial^2 u}{\partial y^2}\right]_{y=0} + \left[\frac{\partial \mu}{\partial y} \frac{\partial u}{\partial y}\right]_{y=0} = -\rho_\delta u_\delta \frac{du_\delta}{dx}. \tag{3.63}$$

For constant molecular viscosity μ (e.g. in incompressible flow without heat transfer), one has $(\partial \mu/\partial y)_{y=0} = 0$. In this case, eq. (3.63) simplifies to the important statement, already discussed earlier, that at the wall $y = 0$ the second derivative of the velocity profile with respect to the distance y from the wall is proportional to the pressure gradient dp/dx, or the velocity gradient du_δ/dx in the direction of flow.

In the case of *retarded* free-stream flow $u_\delta(x)$, that is, for $du_\delta/dx < 0$, $dp/dx > 0$, we have, according to eq. (3.63), $(\partial^2 u/\partial y^2)_{y=0} > 0$. As y approaches δ, the second derivative $\partial^2 u/\partial y^2$ has to be negative, according to eq. (3.60). Thus, $\partial^2 u/\partial y^2$ must have a zero in the interval $0 < y < \delta$; that is, the velocity profile must have an inflection point. The distance y from the wall at which the inflection point occurs depends on the previous history of the boundary layer.

In *accelerated* free-stream flow $u_\delta(x)$, that is for $du_\delta/dx > 0, dp/dx < 0$, we find $(\partial^2 u/\partial y^2)_{y=0} < 0$, and thus $\partial^2 u/\partial y^2$ is negative in the entire interval $0 < y < \delta$ so long as the velocity profile did not already have an inflection point and thereby positive values $\partial^2 u/\partial y^2$ as a result of a pressure increase in its past history.

The more general formulation of the compatibility condition in eq. (3.63) serves to estimate easily in what way a molecular viscosity μ, which varies with the distance y from the wall, influences the laminar velocity profile in the vicinity of the wall. To this end, we rewrite eq. (3.63) as follows:

$$\left[\mu \frac{\partial^2 u}{\partial y^2}\right]_{y=0} = -\rho_\delta u_\delta \frac{du_\delta}{dx} - \left[\frac{\partial \mu}{\partial y} \cdot \frac{\partial u}{\partial y}\right]_{y=0}. \tag{3.64}$$

The influence of the term with the factor $(\partial \mu / \partial y)_{y=0}$ is therefore comparable to that of the velocity gradient du_δ / dx. Obviously, the algebraic sign of $(\partial \mu / \partial y)_{y=0}$ plays an important role [$(\partial u / \partial y)_{y=0}$ is always positive)]. For gases and fluids, the viscosity μ depends practically only upon T, as noted earlier (Sect. 1.2); for gases μ increases with increasing temperature but decreases for fluids. For a given temperature gradient $(\partial T / \partial y)_{y=0} > 0$, which corresponds to a cooling of the wall, $(\partial \mu / \partial y)_{y=0}$ is therefore positive for gases and negative for fluids. In the assumed case $(\partial T / \partial y)_{y=0} > 0$ the variable viscosity for *gases* acts according to eq. (3.64), because $(\partial \mu / \partial y) > 0$, just like a positive velocity gradient (an accelerated external flow) and thus the velocity profiles become more concave as a result of this influence [$\partial^2 u / \partial y^2$ becomes more negative]. For the same case (cooling of the wall) in *fluids*, the influence of the term with the factor $(\partial \mu / \partial y)_{y=0} < 0$ is the same as a negative velocity gradient (a retardation of the external flow). The velocity profiles therefore become less concave. In the case of flow along a flat plate $(du_\delta / dx = 0)$ one would expect velocity profiles with inflection points if the wall is cooled.

This more qualitative evaluation of the physical meaning of the compatibility condition (3.64) already illustrates the significance of this relation for approximate computations of the boundary layer. Later on, eq. (3.64) is transformed into a dimensionless form which is suitable for such computations (Sect. 4.2).

Besides the compatibility condition (3.62), it is possible to derive arbitrarily many additional such conditions for the point at the wall, $y = 0$, $u = 0$, $v = 0$, by partial differentiation with respect to y of Prandtl's boundary layer eq. (2.20). This process yields, after one differentiation of eq. (2.20) at $y = 0$,

$$\left[\frac{\partial^2}{\partial y^2}\left(\mu \frac{\partial u}{\partial y}\right)\right]_{y=0} = 0. \tag{3.65}$$

For nonconstant viscosity, differentiation of (3.65) leads to

$$\left[\mu \frac{\partial^3 u}{\partial y^3} + 2 \frac{\partial \mu}{\partial y} \frac{\partial^2 u}{\partial y^2} + \frac{\partial^2 \mu}{\partial y^2} \frac{\partial u}{\partial y}\right]_{y=0} = 0. \tag{3.66}$$

For constant viscosity, eq. (3.66) simplifies to the statement

$$\left[\frac{\partial^3 u}{\partial y^3}\right]_{y=0} = 0. \tag{3.67}$$

The conditions (3.66) and (3.67), which are exact consequences of eq. (2.20), can easily be accounted for in the trial solutions for the velocity profile. For the case of nonconstant viscosity, eq. (3.66) leads to a coupling condition for the properties of the velocity and temperature profiles in the vicinity of the wall, which can be used to improve approximate computations.

We shall forego deriving further compatibility conditions from eq. (2.20). In the following section, however, we shall look at a few more compatibility conditions for the temperature boundary layer which can be derived from eq. (2.91) directly or after one additional partial differentiation.

3.4.3 Compatibility conditions for the temperature boundary layer

From eq. (2.91) the following relation follows for $y = 0$, $u = 0$, $v = 0$:

$$\left[\mu \left(\frac{\partial u}{\partial y} \right)^2 \right]_{y=0} = - \left[\frac{\partial}{\partial y} \left(\lambda \frac{\partial T}{\partial y} \right) \right]_{y=0} . \tag{3.68}$$

This relation expresses the fact that the curvature of the temperature function in the vicinity of the wall is determined solely by the dissipation that prevails there. The relation is also valid when the assumptions made in Sections 2.3.4 through 2.3.6 [(Pr = const. or = 1, $dp/dx = 0$, T_w = const. $(dT_w/dx = 0)$] for the exact solvability of the system of eqs. (2.20), (2.21), (2.91) are no longer satisfied; in other words, the connection between flow and temperature layers need not be known in this case. Relation (3.68) may therefore be used within the framework of approximation theories for an improved computation of heat transfer when the influences of dp/dx and dT_w/dx are taken into account.

Partial differentiation of eq. (2.91) allows derivation of any number of additional compatibility conditions as in the case of the flow boundary layer. We give here the general form of such a compatibility condition of "second order" which was used with good success by Dienemann [11] for a laminar boundary layer with constant material properties in the computation of heat transfer under the influence of strong gradients of the wall temperature dT_w/dx, (but neglecting dissipation and the pressure gradient).

In order to derive the general form for this compatibility condition, eq. (2.91) is partially differentiated with respect to y. For $y = 0$, $u = 0$, $v = 0$, it follows from the continuity eq. (2.21) that

$$\left(\frac{\partial(\rho u)}{\partial x} \right)_{y=0} = - \left(\frac{\partial(\rho v)}{\partial y} \right)_{y=0} = 0. \tag{3.69}$$

Using this equation and observing that $\mathrm{Pr} = \mathrm{const.}$, we can give the compatibility condition

$$c_{\mathrm{p}}\left[\rho\,\frac{\partial u}{\partial y}\,\frac{\partial T}{\partial x}\right]_{y=0} = \left\{\frac{\partial u}{\partial y}\,\frac{\mathrm{d}p}{\mathrm{d}x} + 2\mu\,\frac{\partial u}{\partial y}\,\frac{\partial^2 u}{\partial y^2} + \frac{\partial \mu}{\partial y}\left(\frac{\partial u}{\partial y}\right)^2\right\}_{y=0}$$

$$+ \frac{c_{\mathrm{p}}}{\mathrm{Pr}}\left\{\frac{\partial^2 \mu}{\partial y^2}\,\frac{\partial T}{\partial y} + \mu\,\frac{\partial^3 T}{\partial y^3} + 2\,\frac{\partial \mu}{\partial y}\,\frac{\partial^2 T}{\partial y^2}\right\}_{y=0}. \qquad (3.70)$$

With the compatibility condition (3.62), the second term in the first pair of braces can be expressed differently:

$$\left(2\mu\,\frac{\partial u}{\partial y}\,\frac{\partial^2 u}{\partial y^2}\right)_{y=0} = 2\left[\frac{\partial u}{\partial y}\,\frac{\mathrm{d}p}{\mathrm{d}x} - \frac{\partial \mu}{\partial y}\left(\frac{\partial u}{\partial y}\right)^2\right]_{y=0}. \qquad (3.71)$$

Thus, eq. (3.70) may be written as follows:

$$c_{\mathrm{p}}\left[\rho\,\frac{\partial u}{\partial y}\,\frac{\partial T}{\partial x}\right]_{y=0} = c_{\mathrm{p}}\left(\rho\,\frac{\partial u}{\partial y}\right)_{y=0}\cdot\frac{\mathrm{d}T_{\mathrm{w}}}{\mathrm{d}x}$$

$$= 3\,\frac{\mathrm{d}p}{\mathrm{d}x}\left(\frac{\partial u}{\partial y}\right)_{y=0} - \left[\frac{\partial \mu}{\partial y}\left(\frac{\partial u}{\partial y}\right)^2\right]_{y=0}$$

$$+ \frac{c_{\mathrm{p}}}{\mathrm{Pr}}\left\{\frac{\partial^2 \mu}{\partial y^2}\,\frac{\partial T}{\partial y} + \mu\,\frac{\partial^3 T}{\partial y^3} + 2\,\frac{\partial \mu}{\partial y}\,\frac{\partial^2 T}{\partial y^2}\right\}_{y=0}. \qquad (3.72)$$

For constant viscosity μ, and constant density ρ, i.e., for $\partial\mu/\partial y = \partial^2\mu/\partial y^2 = 0$, and for $\mathrm{d}p/\mathrm{d}x = 0$, this equation takes the following form used by Dienemann:

$$\left(\frac{\lambda}{\rho c_{\mathrm{p}}}\,\frac{\partial^3 T}{\partial y^3}\right)_{y=0} = \left(\frac{\partial u}{\partial y}\right)_{y=0}\cdot\frac{\mathrm{d}T_{\mathrm{w}}}{\mathrm{d}x}. \qquad (3.73)$$

In the more general form (3.72), the pressure gradient $\mathrm{d}p/\mathrm{d}x$ occurs in addition to $\mathrm{d}T_{\mathrm{w}}/\mathrm{d}x$. In approximation theories for laminar boundary layers the compatibility condition (3.72) may therefore either serve alone or, e.g. in connection with the compatibility conditions (3.64) and (3.65), to describe the influence of the wall temperature gradient $\mathrm{d}T_{\mathrm{w}}/\mathrm{d}x$ and the pressure gradient $\mathrm{d}p/\mathrm{d}x$. This results in an improvement (modification) of the coupling law between velocity and temperature profiles (the modification bringing in x-dependent terms) which may be important, especially in computations of the heat transfer in laminar boundary layers (cf. [33], [142]).

In turbulent boundary layers, the modification of the coupling law (2.156) for the purpose of including the influence of $\mathrm{d}T_{\mathrm{w}}/\mathrm{d}x$ and $\mathrm{d}p/\mathrm{d}x$

by an additional integral condition, is suitably affected by eq. (3.57) (cf. Section 5.3). This equation is of course also valid for laminar boundary layers.

3.4.4 Application of the compatibility conditions in turbulent boundary layers

For the turbulent boundary layer the compatibility conditions presented here are generally of little significance. Possible exceptions are the computational methods of Buri [5], Stratford [128], and Townsend [137], which are based on the compatibility condition (3.62). These methods are used for estimates only (cf. Section 4.2) or in special cases for simplified problems, as for example flows of the type $u_\delta \sim x^m$ ("equilibrium boundary layers,' cf. Sections 3.7.2, 6.3.3).

3.5 Trial Solutions for the Velocity Profile

3.5.1 General remarks

The trial solutions for laminar and turbulent velocity profiles have to be structured such that the greatest number possible of the boundary conditions discussed in Section 3.4 are satisfied. According to Section 1.5, it appears quite permissible for approximate solutions to prescribe the "outer" boundary conditions at the finite distance $y = \delta$ from the wall, since the integral expressions are practically unaffected when $y \geq \delta$ in our approximation theory.

For *laminar* boundary layers it is reasonable to require in addition that the trial solution come as close as possible to known special exact solutions [e.g., the solutions for the flow type $u_\delta(x) \sim x^m$] but in addition have at least one, or better two or more, free parameters in order to provide the freedom to fit the profiles to any given general problem.

In *turbulent* boundary layers ample experimental data may be substituted for unavailable exact solutions. These data are available, critically evaluated and classified according to well-developed semiempirical theories on turbulent exchange phenomena. A short description of the present state of this theory is given in Section 3.6, highlighting the facts that are of particular interest within the context of this book. For additional information we refer to the detailed presentations in [17] and [108].

When developing the trial solutions for the velocity profile we need not be concerned with compressibility and heat transfer, i.e., the parameters M_δ and Θ. The influence of these parameters on the development of the velocity profile is automatically accounted for in the solution of the system of equations, which consists of the integral conditions and the coupling law (2.156). The influence of these parameters can, however, also be accounted for by a coordinate transformation, as proposed for example by Dorodnitsyn [13], Howarth [48], and Geropp [33] (for Geropp's method, cf. Section 4.4).

The same holds as a good approximation for the turbulent boundary layer as well. Strictly speaking, an additional parameter, namely the friction coefficient c_f [for definition, cf. eq. (3.40)] or a Reynolds number based on the boundary layer thickness, appears as shape parameter of the velocity profile. The integral quantities of the approximation theory, however, remain practically independent of these quantities. Only when the influence of compressibility is very strong (at about $M_\delta > 5$), may these parameters become essential. The application of the approximation theories described in this book should therefore be limited to cases with $0 < M_\delta < 5$.*

The basis of the general approximation theory, which applies both to the laminar and turbulent cases, is given by the integral conditions for momentum and energy, eqs. (3.28) and (3.46). These equations suggest using the following quantities for the characterization of the boundary layer: the momentum-loss thickness δ_2, from eq. (3.24), the Reynolds number

$$R_{\delta_2} = \frac{\rho_\delta u_\delta \, \delta_2}{\mu_w}, \tag{3.74}$$

and the ratio

$$\frac{\int_0^\delta \frac{u}{u_\delta}\left[1 - \left(\frac{u}{u_\delta}\right)^2\right] dy}{\int_0^\delta \frac{u}{u_\delta}\left(1 - \frac{u}{u_\delta}\right) dy} = \frac{(\delta_3)_u}{(\delta_2)_u} = H_{32} \equiv H \tag{3.75}$$

as the shape parameter for the velocity profile. Instead of H, the ratio

$$\frac{\int_0^\delta \left(1 - \frac{u}{u_\delta}\right) dy}{\int_0^\delta \frac{u}{u_\delta}\left(1 - \frac{u}{u_\delta}\right) dy} = \frac{(\delta_1)_u}{(\delta_2)_u} = H_{12} \tag{3.76}$$

is quite often used as a shape parameter.

In the case of compressible flow the integral conditions contain not only the characteristic shape-related quantities $(\delta_1)_u$, $(\delta_2)_u$, $(\delta_3)_u$ of the

*Other assumptions, such as the constancy of the specific heats c_p, c_v and their ratio κ may lose their validity in the region $M_\delta > 5$.

velocity profile, but also the physical quantities δ_1, δ_2, δ_3, defined by eqs. (3.23), (3.24), and (3.44), and the ratios δ_1/δ_2 and δ_3/δ_2. The latter ratios should not be confused with H_{12} and H_{32} from eqs. (3.75) and (3.76). Bearing in mind eqs. (2.156) and (2.161), we can write

$$\frac{\delta_1}{\delta_2} = \frac{\delta_1}{\delta_2}(H, M_\delta, \Theta),\tag{3.77}$$

$$\frac{\delta_3}{\delta_2} = \frac{\delta_3}{\delta_2}(H, M_\delta, \Theta).\tag{3.78}$$

In the case of incompressible flow, because $\rho/\rho_\delta = 1$, only the quantities H and H_{12} appear, as defined by eqs. (3.75) and (3.76).

3.5.2 Laminar velocity profiles

3.5.2.1 Trial solutions with one shape parameter. The general form for the trial solutions with *one* shape parameter $m(x)$ and boundary layer thickness $\delta(x)$ as the second characteristic parameter of the boundary layer is, according to Section 3.1,

$$\frac{u(x, y)}{u_\delta(x)} = f\left[\frac{y}{\delta(x)}, m(x)\right].\tag{3.79}$$

It is suggested to represent the function f in (3.79) by a polynomial in

$$\frac{y}{\delta(x)} = \eta(x, y).\tag{3.80}$$

Based on investigations by Mangler [69], we can conveniently put

$$\frac{u}{u_\delta} = 1 - (1 - \eta)^\varsigma[1 + a_1\eta + a_2\eta^2 + a_3\eta^3 + \cdots].\tag{3.81}$$

This expression automatically satisfies the boundary conditions $\eta = 0$, $u/u_\delta = 0$, and $\eta = 1$, $u/u_\delta = 1$. The choice of the exponent ς determines how many derivatives of u/u_δ with respect to η vanish at the point $\eta = 1$. If all derivatives up to nth order [see eq. (3.60)] are to vanish, ς has to be chosen greater than n. The coefficients of the polynomial in the brackets in eq. (3.81) are then determined (at least for incompressible flow) so as to satisfy "compatibility conditions," according to Section 3.4.2.* One of

*In compressible boundary layers the coefficients in eq. (3.81) become dependent upon the Mach number M_δ and the parameter Θ for heat transfer as a result of satisfying the compatibility conditions. The velocity profiles would turn out to be dependent on three parameters in this case, because of the two additional parameters M_δ and Θ. In approximation theories for compressible boundary layers with only one shape parameter, usually no attempt is made to satisfy the compatibility conditions (cf. Section 4.3).

these coefficients, however, has to play the role of the shape parameter for the velocity profile.

Pohlhausen [91] in his well-known computational method for incompressible laminar boundary layers chose $\zeta = 3$, thus satisfying the boundary condition (3.60) at $\eta = 1$ up to second order:

$$\left(\frac{\partial u/u_\delta}{\partial \eta}\right)_{\eta=1} = \left(\frac{\partial^2 u/u_\delta}{\partial \eta^2}\right)_{\eta=1} = 0. \tag{3.82}$$

Moreover, the compatibility condition (3.62) is satisfied in the case of constant material properties ($\mu = $ const., $\rho = $ const.). It is therefore suggested to introduce the second derivative of the velocity with respect to the distance from the wall, η,

$$-\left(\frac{\partial^2 u/u_\delta}{\partial \eta^2}\right)_{\eta=0} = \gamma, \tag{3.83}$$

as the shape parameter of the velocity profile. The general eq. (3.81) simplifies in the case of the Pohlhausen method with a single shape parameter γ to

$$\frac{u}{u_\delta} = 1 - (1 - \eta)^3(1 + a_1\eta), \tag{3.84}$$

where

$$a_1 = 1 - \frac{\gamma}{6}. \tag{3.85}$$

Figure 3.3 shows the family of one-parametric velocity profiles according to eq. (3.84) with γ as the shape parameter.

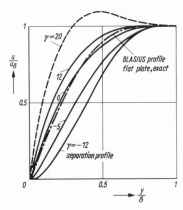

Fig. 3.3. Velocity profiles according to Pohlhausen [91] based on a fourth-degree polynomial, eq. (3.84), in $\eta = y/\delta$ with the shape parameter γ from eq. (3.83).

The "separation profile" with $[\partial(u/u_\delta)/\partial\eta]_{\eta=0} = 0$, which may occur in strongly retarded flows is characterized by the value $\gamma = -12$. For values of $\gamma > 12$, eq. (3.84) yields velocity profiles with $u > u_\delta$ ($u/u_\delta > 1$), which are normally without physical meaning (see Fig. 3.3).

Velocity profiles with values of γ greater than 12, though without "excess velocities," are actually observed as for example in very strongly accelerated flows and in the case of boundary layer suction. This phenomenon demonstrates an apparent limitation to the usefulness of the rather simple eq. (3.84). There is no basic difficulty, however, in avoiding the purely formal mathematical deficiencies of the trial solution (3.84). According to the remarks in Section 3.5.1, it is useful to adopt the one-parametric family of Hartree's velocity profiles (cf. Table 3.1, pp. 77–79), the exact solutions for flows of the type $u_\delta(x) \sim x^m$, as trial solutions in the approximation theory for arbitrary flow, and to represent these by polynomials of the type (3.81). This family does not contain any velocity profiles with excess velocities. According to Geropp [33], it is important in this approximation to assume the exponent ζ in eq. (3.81) to be a function of the shape parameter m, $\zeta = \zeta(m)$. Moreover, it is necessary to carry terms up to the third order in the bracket of eq. (3.81) that is, one must retain the coefficients $a_1(m)$, $a_2(m)$, and $a_3(m)$. Instead of the shape parameter m, it is of course also possible to introduce the Pohlhausen parameter $\gamma = \gamma(m)$ in the definition (3.83) [the trial solution (3.81) depends here on one parameter only], or any other quantity characteristic of the velocity profile (such as H from eq. (3.75)). A very good approximation of the Hartree profiles can be achieved by the following analytical expressions for a_1, a_2, a_3, ϵ, and ζ as functions of the parameter γ:

$$a_1 = \zeta - \epsilon, \tag{3.86}$$

$$a_2 = \frac{\gamma}{2} - \zeta\epsilon + \frac{\zeta + \zeta^2}{2}, \tag{3.87}$$

$$a_3 = \frac{\gamma}{2}\zeta - \epsilon\frac{\zeta + \zeta^2}{2} + \frac{2\zeta + 3\zeta^2 + \zeta^3}{6}, \tag{3.88}$$

$$\epsilon = \left(\frac{\partial u/u_\delta}{\partial \eta}\right)_{\eta=0}, \tag{3.89}$$

$$\epsilon = -0.89855 + \sqrt{1.60901\gamma + 13.6795}, \tag{3.89a}$$

valid for one-parametric Hartree profiles

$$\zeta = 5.22550 + \sqrt{1.30839\gamma + 10.85171}. \tag{3.90}$$

Here we have

$$\gamma(m) = 38.745\,\beta^* - 7.1178\,\beta^{*\,2} + 6.3726\,\beta^{*\,3}, \tag{3.91}$$

$$\beta^* = \frac{2m}{m+1}. \tag{3.92}$$

Sample computations show (somewhat surprisingly) that different one-parametric solutions, such as (3.84), (3.85), and (3.81) together with (3.86) through (3.92) lead to almost equally good results if the computations are based on the integral conditions for momentum and energy (cf. Section 6.4.1, Fig. 6.12a).

Here follow some indications of the range of values of the shape parameter H, as given by eq. (3.75), in view of the computational method to be developed later. The separation profile, characterized by vanishing shear stress at the wall, corresponds to $H = 1.515$, according to Hartree (see Table 3.1). The most concave profile in the Hartree class, which occurs in accelerated flow with $\beta^* \to \infty$, is characterized by $H \approx 1.65$.

The most concave velocity profile imaginable by physical intuition is the rectangular profile. In order to be able to determine the corresponding shape parameter H, a trial function is chosen for the velocity profile. The trial function $u/u_\delta = 1 - (1 - \eta)^\xi$, a simplified form of the general expression (3.81), represents the rectangular profile in the limit $\xi \to \infty$. The corresponding limit for H is $\frac{5}{3} \approx 1.667$. The choice of another trial function for the velocity profile may produce other limiting values for H as the rectangular profile is approached, as for example $H = 2.0$ for the power law in eq. (3.123). Eppler's conclusion [23] that the rectangular profile always corresponds to $H = 2.0$, regardless of the trial function chosen for u/u_δ, is not valid.

3.5.2.2 Trial solutions containing two shape parameters. It is clear that the possibility to approximate arbitrary velocity profiles, and thus the overall accuracy of an approximation theory, can be improved if two or more free, i.e., x-dependent, shape parameters are provided for in the trial solution. In this section, a two-parametric polynomial trial solution is described, which yields extraordinarily accurate solutions for the laminar boundary layer problem when used with a suitable system of three equations.

A trial solution of the type (3.81) with two shape parameters was already used by Wieghardt [155]. He put

$$\frac{u}{u_\delta} = 1 - (1 - \eta)^8[1 + a_1\eta + a_2\eta^2 + a_3\eta^3], \tag{3.93}$$

that is, he chose a polynomial of eleventh degree in η with a fixed value of $\zeta = 8$, thereby satisfying the outer boundary conditions at $\eta = 1$ up to the seventh order. For shape parameters he chose the quantities

$$\alpha = \left[\frac{\partial(u/u_\delta)}{\partial(y/(\delta_2)_\mathrm{u})} \right]_{y=0} = \epsilon \frac{(\delta_2)_\mathrm{u}}{\delta}, \tag{3.94}$$

$$(\delta_2)_\mathrm{u} = (\delta_2)_{\frac{\rho}{\rho_\delta}=1} = \int_0^\delta \frac{u}{u_\delta}\left(1 - \frac{u}{u_\delta}\right) dy, \tag{3.95}$$

[see also footnote concerning $(\delta_1)_\mathrm{u}$ on p. 22] and

$$\Gamma = - \left[\frac{\partial^2(u/u_\delta)}{\partial[y/(\delta_2)_\mathrm{u}]^2} \right]_{y=0} = \gamma \left[\frac{(\delta_2)_\mathrm{u}}{\delta} \right]^2, \tag{3.96}$$

where ϵ is given by eq. (3.89) and γ by eq. (3.83).

The coefficients a_1, a_2, and a_3 of the trial solution (3.93) and all universal expressions occurring in the integral conditions for momentum and energy can be expressed as functions of α and Γ by very elementary computations. It turns out that not every arbitrary pair α, Γ leads to physically meaningful velocity profiles (with monotone increasing u/u_δ between $0 < \eta < 1$ without assuming values $u/u_\delta > 1$). For values of $\alpha > 0.371$, which occur in accelerated flows, as for example in the vicinity of a stagnation point, the trial solution (3.93) fails and is therefore less valid in this domain than the one-parametric expression (3.84) given by Pohlhausen [91] (cf. Fig. 3.3). Wieghardt therefore used that trial solution only for the boundary layer computation in retarded flows and achieved considerable improvements in the prediction of the separation point over the one-parametric computations.

Head [44] proposed a modification of the trial solution (3.93) that defies analytic description but that permits avoiding the aforementioned difficulties within a limited range of accelerated flows.

A different way to avoid, in principle, the velocity profiles with $u/u_\delta > 1$ in a two-parametric trial solution of type (3.81) was pointed out by Geropp [33]. He chose one of the two shape parameters to be the exponent ζ in eq. (3.81). Based on ideas of Mangler [69], a certain coupling between ζ and ϵ is prescribed such that ζ increases with increasing ϵ and that the coupling law $\zeta(\epsilon)$ corresponds reasonably well with the result for special exact solutions (for example, the "similar" solutions by Hartree [43], cf. Section 2.2.1.2).

Geropp's two-parametric trial solution is given by

$$\frac{u}{u_\delta} = 1 - (1 - \eta)^\zeta[1 + a_1(\zeta, \gamma)\eta + a_2(\zeta, \gamma)\eta^2 + a_3(\zeta, \gamma)\eta^3], \tag{3.97}$$

where a_1, a_2, a_3 are given by eqs. (3.86) through (3.88) and a coupling relation between ζ and ϵ:

$$\zeta(\epsilon) = 7 + 1.7513\epsilon - 0.7026\epsilon^2 \qquad (0 < \epsilon < 1); \qquad (3.98)$$

$$\zeta(\epsilon) = 8 - 0.0235\epsilon + 0.0722\epsilon^2 \qquad (\epsilon > 1). \qquad (3.99)$$

A shape parameter that is more convenient for practical computations than ζ will be introduced later [cf. Section 4.4, eq. (4.87)].

Geropp computed all universal functions that occur in the integral conditions for momentum and energy and developed a simple scheme for the computation of laminar boundary layers, valid also in compressible flow [see Computational Scheme B (p. 276)]. A discussion of his method will be given in Section 4.4.

3.5.3 Turbulent velocity profiles

3.5.3.1 General remarks. Section 1.3 dealt with the origin of shear stresses as a consequence of the turbulent exchange of momentum between domains of different velocities in the main flow. These shear stresses determine, through their interaction with the other forces (inertial and pressure forces), the development of the velocity component u parallel to the wall (its time average) as a function of the distance, y, from the wall, in other words the velocity profile.

Whereas the shear stresses in the laminar boundary layer are connected with the molecular viscosity by Newton's law, the apparent viscosity in turbulent boundary layers can only be described empirically (at the present time). It is indeed possible to use suitable models for the turbulent mechanism, such as Prandtl's mixing path (cf. Section 1.3) and similarity hypotheses, to reduce the number of empirical constants or relations. The theory of turbulent flow rests largely on an empirical basis till today and leaves many important questions unanswered. For a detailed account of the present status of research in turbulence, we refer to the pertinent literature [72], [108], [111]. It seems important, however, to note here that the empirical laws available today, together with the integral conditions for momentum and energy derived in Section 3.2, are sufficient to predict the development of turbulent boundary layers with adequate accuracy. The data referred to are universal measurements for turbulent velocity profiles, empirical laws for the shear stress τ_w, and dissipation in turbulent boundary layers. The last-mentioned laws, improved and generalized in recent times, will be the topics of Sections 3.6 and 3.7.

3.5.3.2 The laminar sublayer. First an important concept should be mentioned which is a necessary consequence of Prandtl's model of the turbulence mechanism (Section 3.3): The laminar sublayer. As the distance from the wall tends to zero ($y \rightarrow 0$), the mixing path l_t also goes to zero, according to eq. (1.19). Thus, as $y \rightarrow 0$, the velocities $\tilde{u}$ and $\tilde{v}$ of the turbulent exchange motion also vanish. The boundary layer therefore becomes again laminar in the immediate vicinity of the wall.

The thickness, δ_l, of this laminar sublayer is assumed to be such that for distances from the wall $y > \delta_l$, the full turbulent boundary layer is suddenly present. Actually, between the purely laminar sublayer and the fully turbulent outer layer there is a buffer layer which from time to time may change from laminar to turbulent state and vice versa.

This simplifying assumption of a purely laminar and a fully turbulent boundary layer domain with a sharp, fictitious boundary between the two domains appears justified in approximation theories (at least for the evaluation of the integral expressions in these theories), if one allows for the fact that the laminar sublayer together with the intermediate layer is generally (for large Reynolds numbers) only a small fraction of the total boundary layer thickness. In compressible boundary layers, however, one finds relatively thick laminar sublayers (up to several percent of the total boundary layer thickness for Mach numbers exceeding 5, provided the wall is not cooled). The increase in the thickness of the laminar sublayer with rising Mach number is explained by an increase in the molecular viscosity with the temperature which dampens turbulence in this domain near the wall.

The velocity distribution inside the laminar sublayer, in a first approximation, may be assumed linear because of its small thickness:

$$\frac{u}{u_\delta} = \epsilon\eta \quad (0 < y < \delta_l; \; 0 < u < u_l), \tag{3.100}$$

where ϵ is given by eq. (3.89a). As a consequence of the linear dependence of u on y, we find for the laminar sublayer that

$$\epsilon = \frac{u_l}{\delta_l}\frac{\delta}{u_\delta} = \frac{\tau_w}{\mu_w}\frac{\delta}{u_\delta} = \frac{\tau_w}{\rho_\delta u_\delta^2}\frac{\rho_\delta u_\delta \delta}{\mu_w}. \tag{3.101}$$

The tangent at the wall, ϵ, is generally obtained from the empirically determinable shear stress at the wall, τ_w, because of the difficulties connected with a direct measurement of the velocity for small distances from the wall. (This will be discussed in Section 3.6.) The linear relation (3.101) allows for a simple definition of the thickness δ_l of the laminar sublayer, namely, the distance from the wall at which the straight line given by (3.101) intersects the graph of u/u_δ, valid for the fully turbulent domain.

In a more refined theory, one might ask that at the distance $y = \delta_l$

not only the velocities u/u_δ agree, but also the derivatives $\partial(u/u_\delta)/\partial\eta$ of the purely laminar and fully turbulent domain. For purposes of an approximation theory based on integral conditions, the approximation (3.100) suffices in connection with a trial solution for the fully turbulent domain which will be treated now.

3.5.3.3 The fully turbulent domain. (*a*) *The vicinity of the wall:* the character of the velocity distribution in the domain that is fully turbulent but still near the wall can be determined, according to Prandtl [95], from the simple model of the mixing path l_t. To this end, we assume tentatively that the shear stress $|\tau| = \tau_w$, eq. (1.18), be constant. That equation, together with eq. (1.19), then becomes

$$\frac{\tau_w}{\rho} = c_1^2 y^2 \left(\frac{\partial u}{\partial y}\right)^2 \quad (\delta_l < y < \delta) \tag{3.102}$$

or

$$\frac{1}{c_1 y}\sqrt{\frac{\tau_w}{\rho}} = \frac{\partial u}{\partial y}. \tag{3.103}$$

For constant density ρ, integration of eq. (3.103) yields

$$u(y) = \frac{\sqrt{\tau_w/\rho}}{c_1} \ln y + c_2. \tag{3.104}$$

The integration constant c_2 can be determined from the condition that at the distance $y = \delta_l$ (the inner edge of the fully turbulent domain) the velocity is $u = u_l$. Thus we have

$$c_2 = u_l - \frac{\sqrt{\tau_w/\rho}}{c_1} \ln \delta_l. \tag{3.105}$$

Substituting this expression for c_2 in eq. (3.104) yields

$$u(y) = \frac{\sqrt{\tau_w/\rho}}{c_1} \ln\left(\frac{y}{\delta_l}\right) + u_l = \sqrt{\frac{\tau_w}{\rho}}\left[\frac{1}{c_1}\ln\left(\frac{y}{\delta_l}\right) + \frac{u_l}{\sqrt{\tau_w/\rho}}\right]. \tag{3.106}$$

Equation (3.106) suggests introducing the expression

$$\sqrt{\frac{\tau_w}{\rho}} = u_\tau \tag{3.107}$$

with the dimension of a velocity, the so-called "shear stress velocity" u_τ, as a measure for the velocity u. One may then write eq. (3.106) in the form

$$\frac{u}{u_\tau} = \frac{1}{c_1} \ln \frac{y}{\delta_l} + \frac{u_l}{u_\tau} \,. \tag{3.108}$$

Because of the linear velocity distribution in the laminar sublayer, as described by eqs. (3.100) and (3.101), the thickness δ_l of that layer can be expressed as follows:

$$\delta_l = \mu \frac{u_l}{\tau_w} = \mu \frac{u_l}{u_\tau} \cdot \frac{1}{\rho u_\tau} \,. \tag{3.109}$$

Using eq. (3.109), one may write for eq. (3.108)

$$\frac{u}{u_\tau} = \frac{1}{c_1} \ln \left(\frac{\rho u_\tau y}{\mu} \cdot \frac{u_\tau}{u_l} \right) + \frac{u_l}{u_\tau} = \frac{1}{c_1} \ln \frac{\rho u_\tau y}{\mu} + \left[\frac{1}{c_1} \ln \frac{u_\tau}{u_l} + \frac{u_l}{u_\tau} \right] \,. \tag{3.110}$$

Extensive experimental results show that the ratio u_l/u_τ, and thus the entire expression in the brackets on the right-hand side of eq. (3.110), is a universal constant. We may therefore write eq. (3.110) as the "universal law of the wall"

$$\frac{u}{u_\tau} = \frac{1}{c_1} \ln \left(\frac{\rho u_\tau y}{\mu} \right) + c_3, \tag{3.111}$$

where

$$c_1 = 0.40; \quad c_3 = 5.1.^* \tag{3.111a}$$

The expression (3.111) is only valid in the domain $\delta_l < y \ll \delta$. [For $y = 0$, eq. (3.111) would yield the physically meaningless value $u = -\infty$.] By using the dimensionless distance from the wall

$$\eta^* = \frac{\rho u_\tau y}{\mu} = \frac{y}{\mu/\rho u_\tau} \,, \tag{3.112}$$

which assumes the form of a Reynolds number based on the distance from the wall, y, and the shear stress velocity u_τ, the relation (3.100) for the *laminar sublayer* may also be written as

$$\frac{u}{u_\tau} = \eta^*. \tag{3.113}$$

---

*The values given by different authors for the constant c_3 vary between 5.1 and 5.5. According to very careful investigations carried out by Rechenberg [103], [104] with modern measuring devices, the values $c_3 = 5.1$ and $c_1 = 0.40$ appear most acceptable.

Present research seems to indicate that the relations (3.111) through (3.113) are also valid for compressible flow (at least approximately), if the material constants μ and ρ that prevail in the vicinity of the wall ($\mu \approx \mu_w$, $\rho \approx \rho_w$) are used.

(*b*) *The outer domain:* The relation (3.111) was derived on the assumption that the shear stress τ outside the laminar sublayer be constant and equal to the shear stress at the wall τ_w. Since at the outer edge $y = \delta$ of the fully turbulent domain of the boundary layer $(\partial u / \partial y) \to 0$, and thus also $\tau \to 0$, eq. (3.111) can only be valid in a limited domain near the wall, $\delta_l < y \ll \delta$. This is already obvious from the occurrence of the logarithmic term in eq. (3.111), which would yield $u \to \infty$ as $y \to \infty$, while actually for $y \to \infty$ (or $y = \delta$) we have to expect $u = u_\delta$.

It is therefore necessary to augment or modify eq. (3.111) in such a way that the outer domain of the fully turbulent boundary layer, which generally dominates, is reproduced in agreement with experimental observations. It turns out that the velocity distribution in this domain depends strongly on the pressure gradient of the external flow $u_\delta(x)$, but can essentially be represented by a one-parametric family of curves (see Fig. 3.4). The requirement of a continuous transition to the expressions valid in the vicinity of the wall results in a dependence, although generally weak, of the turbulent velocity profiles on the drag coefficient:

$$\frac{\tau_w}{\rho_\delta u_\delta^2} = \left(\frac{u_\tau}{u_\delta}\right)^2 = \frac{c_f}{2}. \tag{3.114}$$

Among the numerous proposals made to describe the empirical results in the outer domain of the fully turbulent boundary layer, the papers of Coles [9] merit special mention. According to him, the velocity in the

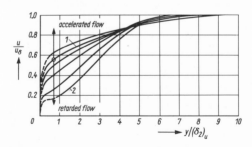

Fig. 3.4. Family of fully turbulent velocity profiles, obtained from measurements by Gruschwitz [41].
1.   flat plate
2.   turbulent separation.

outer domain of the turbulent boundary layer can be represented by a universal law which is also observed to hold in the domain of turbulent mixing of a free jet or in the wake behind a solid body. This universal law is actually independent of the laws (3.111) and (3.113), which depend on the shear stress at the wall τ_w, but it may be combined with those.

According to Coles, the velocity profile for the entire domain of the fully turbulent boundary layer can be described by

$$\frac{u}{u_\tau} = \frac{1}{c_1} \ln \eta^* + c_3 + \frac{1}{c_1} \omega(\eta) f^*; \quad \delta_l < y < \delta; \quad \left(\eta = \frac{y}{\delta} \right),$$

(3.115)

where $\omega(1) = 2$ and

$$f^*(x) = \frac{c_1}{2} \left[\frac{u_\delta}{u_\tau} - \frac{1}{c_1} \ln \frac{\rho u_\tau \delta}{\mu} - c_3 \right].$$

(3.116)

This universal "law of the wake" $\omega(\eta)$ by Coles is represented in Fig. 3.5 and may be approximated very well, according to Hinze [46], by the analytical expression

$$\omega(\eta) = 1 + \sin \frac{\pi}{2} (2\eta - 1).$$

(3.117)

The constants c_1 and c_3 have the same numerical values as in the law (3.111).

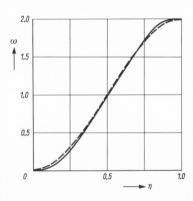

Fig. 3.5. The law of the wake $\omega(\eta)$ by Coles [9]
— — — approximation function

$$\omega(\eta) = 1 + \sin [(\pi/2)(2\eta - 1)]$$

according to Hinze [46].

In order to point out the analogy to the expressions developed in Section 3.5.2 for the laminar boundary layer, we may write (3.115) also in the form

$$\frac{u}{u_\delta} = \frac{u_\tau}{u_\delta} \left\{ \frac{1}{c_1} \ln \eta^* + c_3 + \frac{1}{c_1} \omega(\eta) f^* \right\}; \quad \delta_l < y < \delta.$$

(3.118)

By using the elementary transformation

$$\ln \frac{\rho u_\tau y}{\mu} = \ln \left(\frac{\rho u_\delta \delta}{\mu} \cdot \frac{u_\tau}{u_\delta} \eta \right) = \ln \left(R_\delta \sqrt{\frac{c_f}{2}} \cdot \eta \right) \tag{3.119}$$

and

$$\frac{\rho u_\delta \delta}{\mu} = R_\delta, \tag{3.120}$$

eq. (3.118) may be brought into the form

$$\frac{u}{u_\delta} = F(\eta, R_\delta, c_f); \quad \delta_l < y < \delta. \tag{3.121}$$

For $y = \delta$ eq. (3.118) yields the result $u = u_\delta$, as required.

The expression (3.121) for the turbulent velocity profile thus contains two shape parameters: the local drag coefficient $c_f(x)$ and the Reynolds number R_δ, the latter based on the total boundary layer thickness $\delta(x)$. Equation (3.113) yields a two-parametric representation, also for the *laminar sublayer:*

$$\frac{u}{u_\tau} = \eta^* = \eta \frac{u_\tau}{u_\delta} R_\delta = f(\eta, R_\delta, c_f). \tag{3.122}$$

In eqs. (3.121) and (3.122) one may replace the quantities R_δ and c_f by the parameters H and $R_{\delta 2}$, defined in Section 3.5.1, by applying eqs. (3.24), (3.74), and (3.75). The introduction of these parameters is suggested by the integral conditions of momentum and energy. With reference to the remarks at the end of Section 3.5.3.3a, the relations (3.118) and (3.113) are, at least approximately, valid also for compressible flow.

For approximate computations based on integral conditions, it is often suggested to use the simple power law

$$\frac{u}{u_\delta} = \left[\frac{y}{\delta(x)} \right]^{k(x)} = \eta^{k(x)}; \quad \delta_l < y < \delta; \quad \frac{\delta_l}{\delta} \ll 1 \tag{3.123}$$

for the velocity profiles in the fully turbulent domain, with values for k between 0.1 and 0.7. The approximation of the experimental velocity profiles by this expression, is only acceptable for flows without pressure gradients. For purposes of determining the integral quantities of the approximation theory, only expression (3.118) will, as a rule, be used in the sequel.

Since the laminar sublayer is very thin ($\delta_l / \delta \ll 1$) for large Reynolds numbers, the integral expressions appearing in the approximation theory depend practically only on the fully turbulent part of the boundary layer.

The influence of the laminar sublayer on these integral expressions may only be noticed at very high Mach numbers ($M_\delta > 5$).

For these cases an analytic representation of the fully turbulent domain is desirable which also includes the laminar sublayer.

In the following section such a comprehensive representation will be developed and discussed, a representation which so far has only been tested in the incompressible case by comparison with available measurements.

3.5.3.4 Comprehensive representation for the fully turbulent part including the laminar sublayer. In order to represent the velocity profile fully, it is necessary to piece together, in a suitable manner, the laminar sublayer, eq. (3.113), and the fully turbulent part, eq. (3.115). Such a trial function for the velocity distribution should then, for small η^*, approach eq. (3.113), while for large η^* it should approach eq. (3.115). The expression

$$\frac{u}{u_\tau} = \left[\left(1 - \frac{1}{c_1} - c_3 a \right) \eta^* - c_3 \right] e^{-a\eta^*}$$

$$+ \frac{1}{c_1} \ln(1 + \eta^*) + c_3 + \frac{1}{c_1}\omega(\eta)f^* \qquad (3.124)$$

satisfies these conditions. At the wall, the exact boundary conditions $(u/u_\tau)_w = 0$ and $[(\partial u/u_\tau)/\partial \eta^*]_w = 1$ are satisfied. Equation (3.115), which is only valid in the fully turbulent domain, yields instead the values $-\infty$ or $+\infty$, neither of which is physically meaningful. Figure 3.6 shows

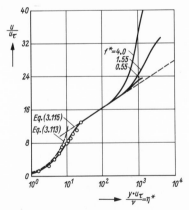

Fig. 3.6.   Universal velocity distribution of the turbulent incompressible boundary layer according to eq. (3.124)
o   Measurements by Reichardt [105].

the velocity distribution in the different domains, with the damping factor a in eq. (3.124) chosen as $a = 0.30$, for the purpose of best fitting the empirical data.

It should be mentioned here that Reichardt [105] already proposed a similar linking of laminar sublayer and turbulent velocity profile near the wall.

Owing to the small thickness of the laminar sublayer, the influence of the parameters c_f and R_{δ_2} on the turbulent velocity profile is very small, with the consequence that, in general, (particularly for incompressible flow) H (or H_{12}) is the only shape parameter necessary. There exists therefore a practically unique connection between H and H_{12}. Numerous older as well as recent measurements were used by Fernholz [29] to establish the following relation, valid for turbulent boundary layers:

$$H_{12} \approx 1 + 1.48(2 - H) + 104(2 - H)^{6.7}. \tag{3.125}$$

In the case of laminar boundary layer with a one-parametric expression, as in eq. (3.81), H_{12} is a single-valued function of H (cf. Appendix I). Moreover, the parameter m, used earlier in eqs. (3.79) through (3.92), is also uniquely coupled with the new parameter H by eq. (3.75) (see Table 3.1).

It should be mentioned here that the wake-function $\omega(\eta)/2$, eq. (3.117), may be considered as the separation profile of the turbulent boundary layer with the tangent at the wall $(\partial u/\partial y)_{y=0} = 0$. For that profile the following holds:

$$\frac{u}{u_\delta} = \tfrac{1}{2}\omega(\eta) \approx \tfrac{1}{2}\left(1 + \sin\left[\frac{\pi}{2}(2\eta - 1)\right]\right). \tag{3.126}$$

It follows from eq. (3.75) that the corresponding shape parameter is exactly $H = 1.500$. As already mentioned at the conclusion of Section 3.5.2.1, the most concave profile is the rectangular one, which is physically obvious. This rectangular profile is obtained from eq. (3.123) for $k = 0$. Based on eq. (3.75) for the definition of H, we find with the velocity distribution (3.123), the general result: $H = 2(1 + 2k)/(1 + 3k)$; thus, for $k = 0$ we have $H = 2$.

The generalization of the expression (3.124) to variable material properties ρ and μ appears possible, in principle. There are, however, some questions which Rotta et al. [109] tried to answer, but their investigations were limited to the relatively small domain of validity of the turbulent profile near the wall, eq. (3.111).

3.6 An Empirical Law for the Shear Stress at the Wall in Turbulent Boundary Layer

The momentum integral condition eq. (3.28) contains the shear stress at the wall, τ_w, in the dimensionless form

$$\frac{2\tau_w}{\rho_\delta u_\delta^2} = c_f. \tag{3.127}$$

One can find through formal manipulations and by observing the definition of the shear stress at the wall, $\tau_w = \mu_w(\partial u/\partial y)_w$ a representation of the local coefficient of friction $c_f = c_f(x)$ which makes apparent the difference between laminar and turbulent boundary layers, both for constant and variable material properties μ and ρ. Without making any restricting assumptions, one may write

$$c_f = 2 \cdot \frac{\mu_w \left(\frac{\partial u}{\partial y}\right)_w}{\rho_\delta u_\delta^2} \cdot \frac{\delta_2}{\delta_2} = 2 \cdot \frac{\left[\frac{\partial(u/u_\delta)}{\partial(y/\delta_2)}\right]_w}{\frac{\rho_\delta u_\delta \delta_2}{\mu_w}} = 2 \cdot \frac{\left[\frac{\partial(u/u_\delta)}{\partial(y/(\delta_2)_u)}\right]_w}{R_{\delta_2}} \cdot \frac{\delta_2}{(\delta_2)_u}. \tag{3.128}$$

R_{δ_2} is the "local" Reynolds number, defined by eq. (3.74) and based on the momentum loss thickness $\delta_2(x)$, eq. (3.24).

In the case of *constant* material properties, such as incompressible flow without temperature differences in the boundary layer, one has $\delta_2 = (\delta_2)_u$, $R_{\delta_2} = R(\delta_2)_u$ and the corresponding local coefficient of friction $c_f = c_{fi}$ is a function of the Reynolds numbers $R_{(\delta_2)_u} = \rho_\delta u_\delta(\delta_2)_u/\mu_w$ for both laminar and turbulent boundary layers and the dimensionless slope at the wall $[\partial(u/u_\delta)/\partial(y/(\delta_2)_u)]_w$. The influence of *variable* material properties, for example in compressible flow without or with heat transfer, is reflected in the factor $\delta_2/(\delta_2)_u$. But there is also an indirect influence through the Reynolds number which contains in its definition the molecular viscosity at the wall, μ_w, the momentum-loss thickness, given by eq. (3.24), and the density ρ_δ.

In a *laminar* boundary layer, where the velocity profiles, both for constant and for variable material properties, are assumed to be characterized by a single shape parameter m or H by eq. (3.81) and eqs. (3.86) through (3.92), [H being defined by eq. (3.75)], the dimensionless slope at the wall is a single-valued function of the shape parameter m or H:

$$\left[\frac{\partial(u/u_\delta)}{\partial(y/(\delta_2)_u)}\right]_w = \alpha_l(m) = \alpha_l(H) \tag{3.129}$$

where $m = \beta^*/(2 - \beta^*)$ from eq. (2.45) and $H = H(\beta^*)$ from Fig. I.1 (p. 260). The function $\alpha_l(H)$ is here given by the trial function in eqs. (3.81), (3.86) through (3.92), while the function

$$\frac{\delta_2}{(\delta_2)_u} = \frac{\delta_2}{(\delta_2)_u}(H, M_\delta, \Theta) = \frac{\int_0^\delta \frac{\rho u}{\rho_\delta u_\delta}\left(1 - \frac{u}{u_\delta}\right)dy}{\int_0^\delta \frac{u}{u_\delta}\left(1 - \frac{u}{u_\delta}\right)dy} \tag{3.130}$$

follows in a unique way from eqs. (3.81), (3.86) through (3.92) together with the coupling relation (2.156) between the velocity and temperature profiles of the boundary layer. Thus, the local coefficient of friction, c_f, is known for the *laminar* boundary layer from eqs. (3.128) and (3.129) in the following form:

$$c_f = c_f(H, M_\delta, \Theta, R_{\delta_2}) = 2\frac{\alpha_l(H)}{R_{\delta_2}} \cdot \frac{\delta_2}{(\delta_2)_u}. \tag{3.131}$$

This relation holds even when the material properties are no longer constant (see also Section 6.2.4).

In the case of *turbulent* boundary layers numerous measurements, based on the assumption of constant material properties, have confirmed the following functional relationship for $c_{fi}(H, R_{(\delta_2)u})$:

$$c_{fi} = 2\frac{\alpha_t(H)}{R_{(\delta_2)u}^n}. \quad * \tag{3.132}$$

This expression was first suggested by Ludwieg and Tillmann [67] and holds for hydraulically smooth surfaces.† The Reynolds number is here defined by

$$R_{(\delta_2)u} = \frac{\rho_\delta u_\delta(\delta_2)_u}{\mu_w} \tag{3.133}$$

($\rho = \rho_\delta =$ const.; $\mu = \mu_w =$ const.) The exponent is $n = 0.268$ and $\alpha_t(H)$ can be represented, according to a critical survey of the literature

---

*It was only later recognized that the expression for c_{fi} is a necessary consequence of the semiempirical trial solution (3.118) for the turbulent velocity profile, if eq. (3.118) is solved for the parameter H, which thus becomes a function of c_{fi} and $R_{(\delta_2)u}$. The resulting expression is then solved for $c_{fi} = c_{fi}(H, R_{(\delta_2)u})$ (cf. for example, Patel [86]).

†A surface is considered hydraulically smooth if the surface roughness $k \leq 100L/R_L$ [$L =$ length of the body, R_L given by eq. (1.39)]. Higher roughness means larger values for c_{fi}. Empirical details on this can be found in Schlichting [111], p. 557.

by Fernholz [29], by

$$\alpha_t(H) = 0.0290 \left[\log_{10} \frac{8.05}{H_{12}^{1.818}} \right]^{1.705}. \tag{3.134}$$

This equation can be approximated quite well by the linear relation in H:

$$\alpha_t(H) \approx 0.0566H - 0.0842. \tag{3.135}$$

For variable material properties, which are encountered for example in compressible flow without or with heat transfer, the general law for $c_f(H, M_\delta, \Theta, R_{\delta_2})$ in turbulent boundary layers can be found in a rather direct way, without the usual simplifying assumptions, by a comparison of eqs. (3.132) and (3.128). For constant material properties this comparison leads at once to the notable result:

$$\left[\frac{\partial(u/u_\delta)}{\partial(y/(\delta_2)_u)} \right]_w = \alpha_t(H) R_{(\delta_2)u}^{1-n}, \tag{3.136}$$

which states that the dimensionless tangent at the wall depends, besides on the shape parameter H, also upon the Reynolds number raised to the power $1 - n = 0.732$, and that it increases in proportion to that power. Equation (3.136) is apparently a generalized form of eq. (3.129): in the special case of the laminar boundary layer it is necessary to put $n = 1$ into eq. (3.136) in order to obtain eq. (3.129).

The general law for c_f for compressible *laminar* boundary layers, eq. (3.131), was obtained by substituting eq. (3.129) into (3.128), where the Reynolds number R_{δ_2} has to be used as defined by eq. (3.120).

The general c_f law for compressible *turbulent* boundary layer can be obtained analogously by substituting eq. (3.136) into (3.128) and replacing $R_{(\delta_2)u}$ by R_{δ_2} from eq. (3.74). This process yields

$$c_f = c_f(H, M_\delta, \Theta, R_{\delta_2}) = 2 \cdot \frac{\alpha_t(H)}{R_{\delta_2}^n} \cdot \frac{\delta_2}{(\delta_2)_u}. \tag{3.137}$$

The function $\alpha_t(H)$ in this expression is identical with the empirically determined law (3.135) for constant material properties. The factor $\delta_2/(\delta_2)_u$ is, as in the case of the laminar boundary layer, a theoretical generalization function, which is given by the expression (3.118) for turbulent boundary layers, and the coupling law (2.156). The function represents the influence of the variable material properties, as they occur, for example, in compressible flow without or with heat transfer. (Analytic expressions for $\delta_2/(\delta_2)_u$, both for laminar and turbulent boundary layers, are found in Appendix I.)

A better understanding of the general law (3.137) is gained by recognizing the physical meaning of eq. (3.136) which is illustrated in Fig. 3.7.

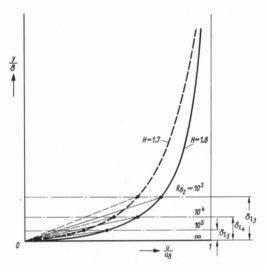

Fig. 3.7.   Illustration of the influence of the Reynolds number upon the thickness δ_l of the laminar sublayer and upon the gradient $(\partial u/\partial y)_{y=0}$ at the wall (schematic only). For the total thickness δ of the boundary layer,

$$\delta \gg \delta_{l_3} > \delta_{l_4} > \delta_{l_5}$$
$$\delta_{l_3} \equiv \delta_l \qquad (R_{\delta_2} = 10^3)$$
$$\delta_{l_4} \equiv \delta_l \qquad (R_{\delta_2} = 10^4)$$

A large Reynolds number R_{δ_2} (or $R_{(\delta_2)u}$ when $\rho = $ const. and $\mu = $ const.) means that the viscosity force proportional to the molecular viscosity μ_w is small compared with the inertial force that exists in fully turbulent flow and is responsible for creating turbulence. The turbulent exchange motions therefore reach closer to the wall for larger values of R_{δ_2} as compared with smaller values of R_{δ_2}. This means that the laminar sublayer decreases in thickness as R_{δ_2} increases. Figure 3.7 shows that for a given succession of velocity profiles in the fully turbulent, outer part of the boundary layer, which determines the shape parameter H and the function $\alpha_t(H)$, there may still be different slopes at the wall, depending on the values of δ_l/δ. The smaller δ_l/δ (i.e., the larger R_{δ_2}), the steeper the slope at the wall (i.e., the larger the angle between tangent and wall) $[\partial(u/u_\delta)/\partial(y/(\delta_2)_u)]_{y=0}$.

Equation (3.136) describes that increase of the slope at the wall as being proportional to $R_{(\delta_2)u}^{1-u}$, in the case of constant material properties.

The assumption that eq. (3.137) holds is now equivalent to postulating that eq. (3.136) holds also for variable material properties, i.e. for compressible flow without and with heat transfer, and that in this case, R_{δ_2} rather than $R_{(\delta_2)u}$ is the only significant parameter for the interaction between laminar sublayer and the turbulent outer part of the boundary layer.

This assumption, which incidentally is the only one made in this generalized theory of the c_f law, appears justified as long as the turbulent exchange mechanism in this medium, which is affected by compressibility and heat transfer, can be considered as an incompressible phenomenon, i.e., as long as the velocity of the turbulent motions does not exceed that of sound.* This assumption is generally well satisfied for Mach numbers up to $M_\delta \approx 5$ in the external flow, an estimate which can easily be obtained from eqs. (1.14), (1.19), and (3.123). [For $k = \frac{1}{7}$, $\eta = 0.5$, for example, one finds $\bar{u}/u_\delta = 0.05$, thus for $M_\delta = u_\delta/a_\delta = 5$, the velocity of the fluctuations $\bar{u}$ is $0.25a_\delta$, i.e., only one quarter of the velocity of sound.]

The ideas developed here are compatible with the results of theoretical and experimental investigations of the same problem by Kovasznay [60], Morkovin [74], and Laufer [64].

3.7 An Empirical Law for Dissipation in Turbulent Boundary Layers

3.7.1 General considerations about the influence of compressibility

The energy-integral condition, eq. (3.46), contains the dissipation integral in dimensionless form:

$$c_D = 2 \cdot \frac{\int_0^{u_\delta} \tau \, du}{\rho_\delta u_\delta^3}. \tag{3.138}$$

This expression can be rewritten, purely formally (for laminar and turbulent boundary layers), as follows:

$$c_D = \frac{2\tau_w}{\rho_\delta u_\delta^2} \int_0^1 \frac{\tau}{\tau_w} \, d\left(\frac{u}{u_\delta}\right) = c_f \int_0^1 \frac{\tau}{\tau_w} \, d\left(\frac{u}{u_\delta}\right). \tag{3.139}$$

---

*If the exchange motions occur with supersonic velocities, one has to expect, among other things, shock waves in the boundary layer and rapidly increasing energy radiation through acoustic waves. These are phenomena that can significantly influence the turbulence mechanism.

For *completely laminar* boundary layers it follows from eq. (3.138) that

$$c_{\mathrm{D}} = 2 \cdot \frac{\displaystyle\int_0^{\delta} \mu \left(\frac{\partial u}{\partial y}\right)^2 \mathrm{d}y}{\rho_{\delta} u_{\delta}^3} = 2 \cdot \frac{\displaystyle\int_0^{\delta/\delta_2} \frac{\mu}{\mu_{\mathrm{w}}} \left(\frac{\partial(u/u_{\delta})}{\partial(y/\delta_2)}\right)^2 \mathrm{d}(y/\delta_2)}{\rho_{\delta} u_{\delta} \,\delta_2/\mu_{\mathrm{w}}}$$

$$= 2 \frac{\beta_l}{\mathrm{R}_{\delta_2}} \cdot \frac{\delta_2}{(\delta_2)_{\mathrm{u}}}, \tag{3.140}$$

where

$$\beta_l = \int_0^{\delta/(\delta_2)_{\mathrm{u}}} \frac{\mu}{\mu_{\mathrm{w}}} \left[\frac{\partial(u/u_{\delta})}{\partial(y/(\delta_2)_{\mathrm{u}})}\right]^2 \mathrm{d}[y/(\delta_2)_{\mathrm{u}}]. \tag{3.141}$$

Since the viscosity μ depends on temperature [cf., for example, eq. (1.7) for air], the function β_l depends not only on the shape parameter H but also on the Mach number and heat transfer. Thus, it is $\beta_l = \beta_l(H, \mathrm{M}_{\delta}, \Theta)$. [Later on, in Section 4.3.3, we shall write $\beta_l = \beta_{\mathrm{u}}(H) \cdot \chi(\mathrm{M}_{\delta}, \Theta)$, where β_{u} depends only on the shape parameter H.]

In turbulent boundary layers the temperature dependence of the molecular viscosity plays a role only in the laminar sublayer. But, as we already showed in Sections 3.5.3.2 and 3.5.3.3, there the shear stress τ is practically constant and equal to τ_{w}, all the way into the domain of the turbulent "law of the wall." In this domain near the wall, τ/τ_{w} is practically equal to unity and thus independent of the Mach number and heat transfer. In the remaining (larger), fully turbulent part the apparent shear stresses prevail which are caused by the turbulent exchange motions, but these stresses are essentially independent of Mach number and heat transfer, as we saw in Section 3.6.

The integral over the shear stresses in eq. (3.139) is therefore practically only a function of the shape parameter H of the velocity profile and the Reynolds number R_{δ_2}, both for compressible and incompressible flows. That function can be developed, just as the function $\alpha_{\mathrm{t}}(H)$ in the law (3.137) for the shear stress at the wall, from measurements in incompressible flow.

3.7.2 Empirical dissipation laws for incompressible flow

Rotta [107], basing his findings on numerous measurements, developed an empirical law for the dissipation c_{Di} in incompressible flow. According to Truckenbrodt [138], this relation can be written as

$$c_{\mathrm{Di}} = \frac{2 \cdot \beta_{\mathrm{t}}}{[\mathrm{R}_{(\delta_2)_{\mathrm{u}}}]^N}, \tag{3.142}$$

where

$$\beta_t = \beta_{u_t} \approx 0.0056 \tag{3.143}$$

and

$$N \approx 0.168 \approx \tfrac{1}{6} \tag{3.144}$$

More recently, Rotta [108] and Fernholz [29] pointed out that eq. (3.142), in conjunction with eqs. (3.143) and (3.144), sometimes fails, particularly for the example of equilibrium boundary layers given by Clauser [8] (similar solutions for the turbulent boundary layer), where the quantities

$$\Pi = \left(\frac{\partial(\tau/\tau_w)}{\partial(y/(\delta_1))_u} \right)_{y=0} = -H_{12} \frac{(\delta_2)_u}{c_{fi}/2} \frac{du_\delta/dx}{u_\delta} * \tag{3.145}$$

$$J = \frac{H_{12} - 1}{H_{12}} \cdot \frac{1}{\sqrt{c_{fi}/2}} \tag{3.146}$$

are held constant. The separation point of the turbulent boundary layer is here computed too far upstream (see Section 6.9). It was deemed important to seek an improvement of the empirical relations (3.142) through (3.144). From the momentum law (3.28) and the energy law (3.46) follows the generally valid expression for c_{Di}:

$$c_{Di} = \frac{c_{fi}}{2} \left\{ \frac{dH}{d(\ln R_{(\delta_2)u})} \left(1 + \frac{1 + H_{12}}{H_{12}} \Pi \right) + H \left(1 + \frac{H_{12} - 1}{H_{12}} \Pi \right) \right\} \cdot \tag{3.147}$$

For $J = $ const., $\Pi = $ const., and $J = J(\Pi)$ we find after elementary transformations, using the empirical laws (3.125) and (3.132),

$$\frac{dH}{d(\ln R_{(\delta_2)u})} = -\frac{n}{2} \frac{f(H)}{df/dH}; \quad f = \frac{H_{12} - 1}{H_{12}} \frac{1}{\sqrt{\alpha_t(H)}} \cdot \tag{3.148}$$

The evaluation of (3.147) and (3.148) shows again that a representation of the form (3.142) is possible with sufficient accuracy, but now

$$\beta_t(H) = 0.00481 + 0.0822(H - 1.5)^{4.81} \tag{3.149}$$

*This relation represents a dimensionless form of the compatibility condition (3.62), which obviously has to be satisfied by the turbulent boundary layer.

$$N(H) = 0.2317H - 0.2644 - 0.87 \cdot 10^5(2 - H)^{20}$$
$$(N = 0 \text{ for } H = 1.500).*$$
(3.150)

Numerical examples in Section 6.9 and also in Walz [151] show that a majority of the chosen cases, which do not correspond to Clauser's equilibrium case, yield the separation point too late if the eqs. (3.149), and (3.150) are used. Equations (3.143) and (3.144), on the other hand, put the separation point generally too far upstream.

A comparison of the free-stream velocity distributions $u_\delta(x)$ in the various examples shows that in the case of equilibrium boundary layers, for which the relations (3.142), (3.149), and (3.150) hold, we have apparently $d^2u_\delta/dx^2 = u_\delta'' > 0$. Rotta's and Truckenbrodt's relations (3.142), (3.143), and (3.144) yield good agreement with the experiment when $u_\delta'' \approx 0$. The dissipation integral c_{Di} apparently depends also on u_δ'', i.e., a dimensionless parameter derived from that quantity, in such a manner that for $u_\delta'' > 0$ an increased turbulent exchange stimulates the energy flow from the free-stream flow to the boundary layer and thus enables the latter to overcome more extended or greater positive pressure gradients. Possibly the effect of longitudinal vortices as observed for example by Fernholz [30] in turbulent pipe flow, is involved here. Such vortices, a possible consequence of the instability of turbulent boundary layers, might be considered carriers and agents of increased transport of kinetic energy from the free-stream flow to the regions near the wall in turbulent boundary layers. Investigations by Betchov and Criminale [2] make it appear possible that an instability of the turbulent boundary layer of Tollmien-Schlichting type (cf. Section 5.1) with transverse vortices, which occurs in free-jet and wake turbulence, might be considered a cause of the increased turbulent exchange motions.

3.7.3 A new dissipation law by K. O. Felsch (extension of Sections 3.7.2 and 6.9)

While the first edition of this book was in press, Felsch [27] succeeded in clarifying to a large extent the question that had been raised about the dissipation law for turbulent boundary layers. If the dissipation integral

---

*Townsend [137] and Stratford [129] investigated an equilibrium profile with vanishing skin friction ($c_f = 0$) which should asymptotically be identical with Coles' wake profile [9], eq. (3.126). Since the dissipation of every wake (or free-jet) profile is independent of the Reynolds number R_{δ_2}, the exponent N in eq. (3.142) should go to zero for that profile. Equation (3.149) yields for $H = 1.5$, with eq. (3.126), the dissipation $c_D = 0.00962$. The same value can be obtained directly from eq. (3.126), if one evaluates eq. (3.138) with the apparent shear stress from eq. (1.18) with a mixing path $l_t = 0.125 \cdot \delta/2$ (which holds for wake and free-jet turbulence, according to Prandtl [95], p. 109.).

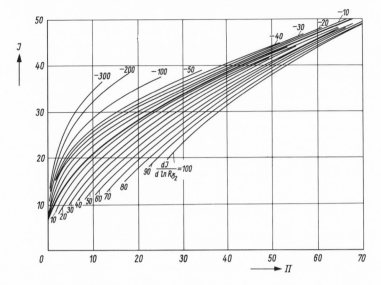

Fig. 3.8.   The function $J = J[\Pi, \mathrm{d}\,J/\mathrm{d}(\ln R_{\delta_2})]$ for turbulent compressible boundary layers ($R_{\delta_2} = R_{(\delta_2)_u}$).

depends, as suspected, on $u_\delta''(x)$, then that influence can also be expressed in terms of the first derivative of the Clauser parameter Π, eq. (3.145), with respect to x, viz. $\mathrm{d}\Pi/\mathrm{d}x \sim u_\delta''$. Since there is a connection (though in general not one-to-one) between Π and the Rotta parameter J, it is possible to take $\mathrm{d}J/\mathrm{d}x$ instead of $\mathrm{d}\Pi/\mathrm{d}x$ as an additional parameter in the dissipation law. Felsch's investigations show that the dimensionless derivative $\mathrm{d}J/\mathrm{d}[\ln R_{(\delta_2)_u}]$ is particularly suited as a parameter for that purpose. He succeeded in establishing a generally valid empirical relation

$$J = J[\Pi; \mathrm{d}J/\mathrm{d}(\ln R_{(\delta_2)_u})], \tag{3.151}$$

which holds for arbitrary turbulent boundary layers (see Fig. 3.8). This relation expresses the physical fact that the shear stress distribution $\tau(y)$ is not uniquely coupled with the velocity distribution $u(y)$, as in the laminar case, but that the functional relation between τ and u contains also the past history of the boundary layer, through the parameter $\mathrm{d}J/\mathrm{d}[\ln R_{(\delta_2)_u}]$. But if this is so, then $\mathrm{d}J/\mathrm{d}[\ln R_{(\delta_2)_u}]$ will also play a role in the integral over the shear stresses, i.e. in the expression for c_{Di}. Using eq. (3.151), one may eliminate the parameter $\mathrm{d}J/\mathrm{d}[\ln R_{(\delta_2)_u}]$; Felsch therefore puts

$$c_{\mathrm{Di}} = c_{\mathrm{Di}}[H_{12}(H), R_{(\delta_2)_u}, \Pi]. \tag{3.152}$$

From eq. (3.147), the definition (3.146) for J, with c_{fi} from eq. (3.132), and the abbreviations

$$J^* = \frac{J}{R_{(\delta_2)u}^{0.134}} = f(H), \tag{3.153}$$

$$J^{*\prime} = \frac{d[f(H)]}{dH}, \tag{3.154}$$

$$J' = \frac{dJ}{d(\ln R_{(\delta_2)u})} = J'(J, \Pi) \tag{3.155}$$

follows, after elementary computations, the relation

$$c_{Di} = \frac{c_{fi}}{2}\left[\left(\frac{1}{R_{(\delta_2)u}^{0.134}}\frac{J'}{J^{*\prime}} - \frac{n}{2}\frac{J^*}{J^{*\prime}}\right)\left(1 + \frac{H_{12}+1}{H_{12}}\Pi\right)\right.$$
$$\left. + H\left(\frac{H_{12}-1}{H_{12}}\Pi + 1\right)\right]. \tag{3.156}$$

The expression in the brackets has to be evaluated step by step with the iteration method II described in Chapter 4 if a numerical example is to be treated.

When this new c_{Di} law is used, the examples discussed in Section 6.9 also yield good agreement with measurements.

3.7.4 The dissipation law for compressible flow

If one assumes, for simplicity's sake, that the form (3.142) of the dissipation law, with β and N given by eqs. (3.143), (3.144) or eqs. (3.149), (3.150), still holds, then a comparison of eqs. (3.139) and (3.142), in the case of incompressible flow, where

$$\frac{\rho}{\rho_\delta} = 1, \quad \frac{\delta_2}{(\delta_2)_u} = 1, \quad R_{\delta_2} = R_{(\delta_2)u},$$

shows that

$$\int_0^1 \frac{\tau}{\tau_w} d\left(\frac{u}{u_\delta}\right) = \frac{\beta_t}{\alpha_t} R_{(\delta_2)u}^{n-N}. \tag{3.157}$$

Equations (3.132) and (3.142) tell that $n - N$ is always positive, and therefore the integral (3.157) increases with $R_{(\delta_2)u}$, though very little, since $0 < n - N < 0.1$.

This functional relation should also hold for compressible flow, according to the arguments developed in Sect. 3.6 (and also within the limits established there). It is only necessary to introduce the Reynolds number R_{δ_2}, with its general definition (3.74). From eqs. (3.142), (3.139), and (3.157) follows then necessarily the *generalized law for dissipation in compressible turbulent boundary layers:*

$$c_D = 2 \frac{\beta_t(H)}{R_{\delta_2}^{N(H)}} \cdot \frac{\delta_2}{(\delta_2)_u} = c_{Di} \frac{\delta_2}{(\delta_2)_u}, \qquad (3.158)$$

where $\beta_t(H)$ and $N(H)$ are given by eqs. (3.143) and (3.144), or by (3.149) and (3.150).

Here again, the universal function $\delta_2/(\delta_2)_u$, eq. (3.130), affects the generalization. If one takes Felsch's new dissipation law, eq. (3.156), as the basis, then c_{Di} in eq. (3.158) has to be taken from eq. (3.156). The question is still open whether eq. (3.151) is altered by the influence of compressibility.

4 Systems of Equations Applicable for Practical Computations

4.1 Preliminary Remarks

The system (3.39) of infinitely many ordinary differential equations, together with the coupling law (2.156) between flow and temperature boundary layers, offers in principle the possibility to determine arbitrarily many x-dependent coefficients of the general trial solution (3.5) for given boundary conditions at the points $y = 0$ and $y = \delta$. As was indicated in Section 3.2, this provides, at least for laminar boundary layer, and so long as the coupling law (2.156) is valid, an accurate method for solving the boundary layer problem, a method whose potential at the present time is not yet fully exhausted.

The investigations by Geropp [33], already referred to, indicate that even two equations of the system (3.39), namely the integral conditions for momentum (3.28) and energy (3.46), together with the compatibility condition (3.62) and the coupling law (2.156), are sufficient to compute *laminar* compressible boundary layers with heat transfer with an accuracy that is entirely adequate for most purposes of practical flow experiments and hydrodynamic research. In some cases it is even possible to ignore the compatibility condition (3.62) or the integral condition for energy (3.46). For *turbulent* boundary layers with the empirical relations of Sections 3.5.3 through 3.5.7 it suffices (in today's view) in every case to compute only with the two integral conditions for momentum and energy.

The equations and the trial solutions of Chapter 3 can be brought into a form that is more useful for practical computations. We choose formally identical notations for laminar and turbulent, incompressible and compressible boundary layers.

Practical experience indicates that a choice between three different computational methods is sufficient for most purposes. In any given case this choice depends on the required accuracy and the problem, and is dictated by the following criteria:

1. *Computational Method I* is based on the integral condition for momentum, eq. (3.28), the compatibility condition, eq. (3.62), and a one-parametric trial solution of the type (3.79). The boundary layer is characterized by two quantities: a shape parameter for the velocity profile and a parameter for the boundary layer thickness. This method is useful for obtaining estimates of laminar and turbulent boundary layers in accelerated or slightly retarded incompressible flow with constant material properties.

2. *Computational Method II* is based on the integral conditions for momentum and energy, eqs. (3.28) and (3.46) and a one-parametric trial solution of the type (3.79). The boundary layer is characterized by the same two quantities used in Method I. The method can be applied to the computation of laminar and turbulent boundary layers in arbitrarily accelerated or retarded incompressible and compressible flows without and with heat transfer (i.e., variable material properties). The accuracy of this method is generally sufficient for purposes of experimental hydrodynamics.

3. *Computational Method III* is based on the integral conditions for momentum and energy, eqs. (3.28) and (3.46) as well as the compatibility condition (3.62), and a two-parametric trial solution of the type (3.97). The boundary layer is here characterized by two shape parameters for the velocity profile and a parameter for the boundary layer thickness. This method is useful for more precise computations of laminar boundary layers in arbitrarily accelerated or retarded incompressible and compressible flows without and with heat transfer (i.e., variable material properties).

The principles of the three computational methods are described in Sections 4.2 through 4.4. Computational details are listed in Appendix I.

4.2 Computational Method I (with Momentum Law and Compatibility Condition)

This computational method corresponds in principle to the suggestion by von Kármán [56] and Pohlhausen [91]; for turbulent boundary layer it follows a proposal by Buri [5]. However, the method is here presented in an improved form following Holstein and Bohlen [47], Tani [132], and Koschmieder and Walz [59]. The (x-dependent) shape parameter for the velocity profile is chosen to be the quantity $\Gamma(x)$, defined by eq. (3.96), i.e. the nondimensional second derivative of the velocity profile at the wall, which is uniquely defined as a function of H. For *incompressible* flow, to which the application of this method should be limited, the momentum-integral condition, eq. (3.28), hereafter called the "momentum law," is as follows:

$$\frac{d(\delta_2)_u}{dx} + (\delta_2)_u[2 + H_{12}]\frac{du_\delta/dx}{u_\delta} - \frac{\tau_w}{\rho_\delta u_\delta^2} = 0, \tag{4.1}$$

where $H_{12} = (\delta_1)_u/(\delta_2)_u = H_{12}(\Gamma)$. For the compatibility condition (3.62) we have, with eq. (2.26) and $\mu = $ const.,

$$\mu\left(\frac{\partial^2 u}{\partial y^2}\right)_{y=0} = \frac{dp}{dx} = -\rho_\delta u_\delta \frac{du_\delta}{dx}. \tag{4.2}$$

If we introduce in eq. (4.1) for the shear stress at the wall $\tau_w/\rho_\delta u_\delta^2$ the notation of eq. (3.132)

$$\frac{\tau_w}{\rho_\delta u_\delta^2} = \frac{\alpha[H(\Gamma)]}{(R_{(\delta_2)_u})^n}, \tag{4.3}$$

then eq. (4.1), because of the factor $(\delta_2)_u^{-n}$ in the term representing the shear stress ($n = 1$ for laminar boundary layer), represents an ordinary differential equation of Bernoulli type for $(\delta_2)_u$. Transformation into an ordinary differential equation of first order is effected by the substitution

$$(\delta_2)_u R_{(\delta_2)_u}^n = Z_u. \tag{4.4}$$

By this substitution, eq. (4.1) becomes

$$\frac{dZ_u}{dx} + Z_u\frac{du_\delta/dx}{u_\delta}[2 + n + (1 + n)H_{12}(\Gamma)] - (n + 1)\alpha(\Gamma) = 0. \tag{4.5}$$

For laminar boundary layers, one has $n = 1$ in eq. (4.3). For turbulent boundary layers, we find $n = 0.268$ from eq. (3.132), as given by Ludwieg and Tillmann [67].

Characterizing the thickness of the boundary layer by the new parameter Z_u, we may write the compatibility condition (4.2) for *laminar* boundary layer ($n = 1$) in the simple form

$$Z_u\frac{du_\delta/dx}{u_\delta} = \Gamma(H). \tag{4.6}$$

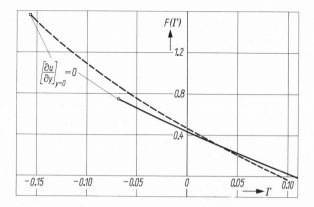

Fig. 4.1. The function $F(\Gamma)$ from eq. (4.8) for the Hartree and Pohlhausen velocity profiles

—— Hartree [43]
– – – Pohlhausen [91], eq. (3.84).

Substitution of eq. (4.6) into (4.5) yields

$$\frac{\mathrm{d}Z_\mathrm{u}}{\mathrm{d}x} = F(\Gamma), \tag{4.7}$$

where

$$F(\Gamma) = 2\alpha_l(\Gamma) - [3 + 2H_{12}(\Gamma)]\Gamma. \tag{4.8}$$

We have, thus, effected a separation of the variables $Z_\mathrm{u}(x)$ and $\Gamma(x)$. In order to solve eq. (4.7), the function $F(\Gamma)$ has to be known. This is the case as soon as one has decided on a specific one-parametric representation for the velocity profile. For laminar boundary layers, Pohlhausen [91] chose the expression (3.84). The relation between γ and Γ is here given by eq. (3.96). According to the introductory remarks about the approximation theory (Section 3.1), it is advantageous to introduce the one-parametric family of exact velocity profiles by Hartree [43] as trial solutions, for example in the form of the interpolation formulas (3.81) and (3.86) through (3.92). The evaluation of the function $F(\Gamma)$ in eq. (4.8) results in the practically linear graph displayed in Fig. 4.1. $F(\Gamma)$ may therefore be represented rather precisely by the following expression:

$$F(\Gamma) \approx a_l - b_l\Gamma = a_l - b_l Z_\mathrm{u} \frac{\mathrm{d}u_\delta/\mathrm{d}x}{u_\delta} \tag{4.9}$$

where

$$a_l = 0.441; \quad b_l = 4.165 \quad \text{for} \quad \Gamma > 0 \quad \text{(accelerated flow)}$$

$$b_l = 5.165 \quad \text{for} \quad \Gamma < 0 \quad \text{(retarded flow)} \qquad (4.10)$$

For the Pohlhausen formula (3.84) we find

$$a_l = 0.470; \quad b_l \approx 5.0 \quad \text{for} \quad \Gamma > 0$$

$$b_l \approx 6.0 \quad \text{for} \quad \Gamma < 0 \qquad (4.11)$$

Using eq. (4.9), we may transform eq. (4.7) into the standard form of an ordinary differential equation of the first order which is immediately integrable:

$$\frac{dZ_u}{dx} + b_l Z_u \frac{du_\delta/dx}{u_\delta} - a_l = 0. \qquad (4.12)$$

The solution of eq. (4.12) is

$$Z_u(x) = Z_{u1} \cdot \left[\frac{(u_\delta)_1}{u_\delta(x)}\right]^{b_l} + a_l \frac{\int_{x_1}^{x} [u_\delta(\xi)]^{b_l}\, d\xi}{[u_\delta(x)]^{b_l}}. \qquad (4.13)$$

On the basis of this equation, the parameter Z_u for the boundary layer thickness may be computed as a function of the distance x for an arbitrary external flow $u_\delta(x)$.

If a location $x = x_1$ is known where the parameter $Z_u = Z_{u1} = 0$ (as is the case at a stagnation point or at the leading edge of the flat plate), eq. (4.13) simplifies to

$$Z_u(x) = a_l \frac{\int_0^x (u_\delta)^{b_l} d\xi}{(u_\delta)^{b_l}}. \qquad (4.14)$$

The integral in (4.13) or (4.14) can be evaluated, either numerically or graphically, for arbitrary free-stream flows $u_\delta(x)$, based on a graph of $[u_\delta(x)]^{b_l}$. In the case when $u_\delta(x)$ is linear between x_1 and x_2,

$$u_\delta(x) = (u_\delta)_1 + c(x - x_1); \quad c = \frac{(u_\delta)_2 - (u_\delta)_1}{x_2 - x_1}, \qquad (4.15)$$

the solution of eq. (4.13) or (4.14) can be represented analytically

$$\frac{Z_u(x)}{Z_{u_1}} = A + B \frac{\Delta x}{Z_{u_1}}; \quad \Delta x = x - x_1, \qquad (4.16)$$

where

$$A = \left[\frac{(u_\delta)_1}{u_\delta}\right]^{b_l};$$ (4.17)

and

$$B = \frac{a_l}{1 + b_l} \cdot \frac{1 - \left[\frac{(u_\delta)_1}{u_\delta}\right]^{1+b_l}}{1 - \frac{(u_\delta)_1}{u_\delta}}.$$ (4.18)

Since an arbitrary function $u_\delta(x)$ can be approximated by a piecewise linear function, the expression (4.16), together with (4.17) and (4.18), is suited to develop the entire solution in a step-by-step fashion, proceeding from point x_{i-1} to point x_i. Best suited for this purpose is a nomogram representation of eq. (4.16) with the scales Z_i/Z_{i-1}, $u_{\delta_i}/u_{\delta_{i-1}}$ and $\Delta x/Z_{i-1}$ (see Fig. 4.2).* A straight line through the given points on the scales $u_{\delta_i}/u_{\delta_{i-1}}$ and $\Delta x/Z_{i-1}$ intersects with the scale Z_i/Z_{i-1}. The desired value Z_i at the end of the interval Δx under consideration is then given by

$$Z_i = \frac{Z_i}{Z_{i-1}} \cdot Z_{i-1}.$$ (4.19)

With $Z(x)$ known, the shape parameter $\Gamma(x)$ follows then immediately from eq. (4.6). Thus, the boundary layer problem is essentially solved. Additional characteristic boundary layer quantities, such as the shear stress at the wall τ_w, the Reynolds number $R_{(\delta_2)_u}$, the momentum-loss thickness $(\delta_2)_u$, and the displacement thickness $(\delta_1)_u$, are immediately computed with the aid of Table 3.1 for $\alpha(H)$, $H_{12}(\Gamma)$, from the following relations

$$\frac{\tau_w}{\rho_\delta u_\delta^2} = \frac{c_f}{2} = \frac{\alpha(\Gamma)}{R_{(\delta_2)u}^n} \qquad (n = 1),$$ (4.20)

$$R_{(\delta_2)u} = \sqrt{Z\frac{\rho_\delta u_\delta}{\mu_w}} = \sqrt{\frac{Z}{L}\frac{\rho_\delta u_\delta L}{\mu_w}} = \sqrt{\frac{Z}{L}}R_L,$$ (4.21)

$$R_L = \frac{\rho_\delta u_\delta L}{\mu_w} = \frac{\rho_\infty u_\delta L}{\mu_\infty}$$ (4.22)

(valid for incompressible flow with $\rho = \rho_\infty = $ const., $\mu = \mu_\infty = $ const.),

$$\frac{(\delta_2)_u}{L} = \sqrt{\frac{Z/L}{R_L}}$$ (4.23)

*The subscript u in Z_u is omitted here, for simplicity's sake. For the case of two-dimensional flow which is treated here, the ratio $\Re_i/\Re_{i-1}$ of the cross-sectional radii of a body of revolution is to be put equal to unity.

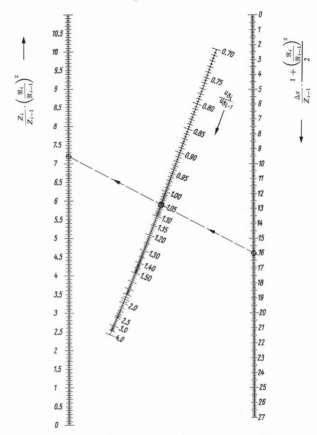

Fig. 4.2. Nomogram for the computation of laminar boundary layers with constant material properties (incompressible flow) based on Method I (see Appendix I.2). Given are: $u_{\delta_i}/u_{\delta_{i-1}}$ and $\Delta x/Z_{i-1}$. To be determined: Z_i/Z_{i-1}; from this follows Z_i with eq. (4.19) and $\Gamma[H(x)]$ with eq. (4.6). For two-dimensional flows the ratio $\mathfrak{R}_i/\mathfrak{R}_{i-1}$ has to be set equal to unity.

(L = reference length, as for example the depth of an airfoil,

$$(\delta_1)_u = H_{12} \cdot (\delta_2)_u, \tag{4.24}$$

$$\delta = \frac{\delta}{(\delta_1)_u} (\delta_1)_u, \tag{4.25}$$

$$\frac{\delta}{(\delta_1)_u} = f(\Gamma) \tag{4.26}$$

(see Fig. I.3 in Appendix I). Separation of the laminar boundary layer occurs when the shear stress at the wall, τ_w, vanishes, i.e., when $\alpha(\Gamma) = 0$.

This happens when $\Gamma = -0.0681$. According to eqs. (3.81) and (3.86) through (3.92), the following relation exists between the shape parameter H, eq. (3.75), and Γ:

$$H \approx 1.572 + \Gamma/1.272 \qquad (4.27)$$

The complete scheme for the computation can be found in Computational Method I in Appendix I.

The computational method described here is only recommended for estimates of laminar and compressible boundary layers. Despite this, the accuracy in many cases is surprisingly high. Theoretical investigations by Nickel [80] are concerned with the limitations of this method which are brought out in the examples of Chapter 6.

For incompressible *turbulent* boundary layer, we put $n = 0.268$ in eq. (4.4), essentially following Buri [5], but we retain the definition (4.6) for Γ. Empirical results suggest putting a_t at 0.0187 and b_t at 4 in eq. (4.9).* According to Buri, separation of the turbulent boundary layer is expected for $\Gamma \leq -0.068$. Between the shape parameter H, which will be used later, and Γ the approximate relation

$$H \approx 1.750 + 2.63\Gamma \qquad (4.28)$$

holds. For the rest, the solutions (4.13) or (4.14) are valid with the above-mentioned numerical values for a_t and b_t. No nomogram is presented here because of the limited reliability of this computational method, but we refer instead to the more precise and only slightly more complicated Method II for turbulent boundary layers, as given in Section 4.3.

4.3 Computational Method II (with Momentum and Energy Laws)

4.3.1 The system of equations

Computational Method II is based on eqs. (3.28) and (3.46). Since it shall also be used for compressible flow with and without heat transfer, it is necessary to observe the general definitions of the integral quantities in eqs. (3.36), (3.37), and (3.38) for $v = 0$ (momentum law) and $v = 1$ (energy law). It is also advantageous, for mathematical reasons, to replace the quantity δ_2 by a quantity Z, in analogy with eq. (4.4):

$$\delta_2 \, R_{\delta_2}^n = Z. \qquad (4.29)$$

*More precisely, it is $b_t = 3.6$ for $\Gamma > 0$, and $b_t = 4.4$ for $\Gamma < 0$. Buri, based on an earlier, less precise, law for the shear stress at the wall, found the values $a_t \approx 0.016$, and $b_t \approx 4$ with $n = 0.25$.

However, here the quantity Z has to be based on the general momentum loss thickness δ_2, given by eq. (3.24) and the general Reynolds number

$$R_{\delta_2} = \frac{\rho_\delta u_\delta \delta_2}{\mu_w}, \tag{4.30}$$

which was introduced in Sect. 3.5.1 [eq. (3.74)]. The exponent n in eq. (4.29) is unity for laminar boundary layers [as a result of the relations (3.131) and (3.137)], and equal to 0.268 for turbulent boundary layers.

On the basis of eq. (4.29), the momentum law (3.28) assumes the following form after elementary computations (' indicates the derivative with respect to x):

$$Z' + Z \frac{u'_\delta}{u_\delta} F_1^* - F_2 = 0, \tag{4.31}$$

where

$$F_1 = 2 + n + (1 + n)\frac{\delta_1}{\delta_2} - M_\delta^2, \tag{4.32}$$

$$F_1^* = F_1 + n \frac{\mu'_w/\mu_w}{u'_\delta/u_\delta} \tag{4.33}$$

$$F_2 = (1 + n)\alpha \frac{\delta_2}{(\delta_2)_u}. \tag{4.34}$$

F_1 and F_2 are universal functions, i.e., they can be evaluated independently of the specific problem as soon as the one-parametric trial solution of type (3.79) (for laminar or turbulent boundary layer) is fixed and the coupling law (2.156) between velocity and temperature profile, together with eq. (2.161), is observed.

The universal functions F_1 and F_2 are therefore dependent on the parameters H, M_δ, and Θ [given by eq. (2.160)]. Analytical expressions for these functions are given in Sect. 4.3.3 and in Appendix I.3 for the laminar boundary layer based on Hartree profiles (Table 3.1), and for turbulent boundary layer based on the empirical law (3.118).

The derivative $\mu'_w = d\mu_w/dx$ of the molecular viscosity at the wall, which occurs in eq. (4.31), has to be considered as given with the problem, just as $u_\delta(x)$, $u'_\delta(x)$, $M_\delta(x)$, $\Theta(x)$.* In many cases, for example for a completely isothermal wall, we have $T_w \approx$ const., and thereby $\mu'_w = 0$ and $F_1^* = F_1$.

---

*When T_w changes rapidly with x, μ_w will do the same, and it should be borne in mind that the coupling law (2.156) may become imprecise. In such cases it is usually sufficient to improve the computation of heat transfer, as shown in Section 5.3, without further iterative improvement of the computation of the flow boundary layer.

It turns out advantageous to introduce into the *energy law* (3.46), instead of the energy-loss thickness δ_3, the ratio

$$\frac{\delta_3}{\delta_2} = H^* \tag{4.35}$$

For incompressible flow, this ratio is identical with the shape parameter H in eq. (3.75):

$$H^*_{\rho/\rho_\delta=1} = \frac{(\delta_3)_u}{(\delta_2)_u} = H. \tag{4.36}$$

In the general case, however, H^* (just as δ_1/δ_2) is a known universal function of H, M_δ, and Θ:

$$H^* = H^*(H, M_\delta, \Theta). \tag{4.37}$$

From (3.46) follows, after elementary computations, and in consideration of the momentum law (4.31), the energy law, with H^* as an auxiliary variable:

$$H^{*\prime} + H^* \frac{u'_\delta}{u_\delta} F_3 - \frac{F_4}{Z} = 0. \tag{4.38}$$

The functions F_3 and F_4 are again universal; they are defined by

$$F_3 = 1 - \frac{\delta_1}{\delta_2} + 2 \frac{\delta_4}{\delta_3}, \tag{4.39}$$

$$F_4 = \frac{c_f}{2} \cdot R_{\delta_2}^n \left(2 \frac{c_D}{c_f} - H^*\right) = \frac{\delta_2}{(\delta_2)_u} (2\beta R_{\delta_2}^{n-N} - \alpha H^*). \tag{4.40}$$

The parameters α and β are given by eqs. (3.129), (3.134), (3.135), (3.141), and (3.143) or (3.149).

For the laminar boundary layer, the function F_4 does not depend on R_{δ_2}, because of $n = N = 1$, $n - N = 0$. For turbulent boundary layers, because of $n - N = 0.268 - 0.168 = 0.1$,[*] there remains a dependence of the function F_4 on R_{δ_2}, though small, whose significance will be discussed later in detail (Sects. 6.2.3, 6.3.3). Analytical expressions for F_3 and F_4 can be found in Appendix I. Separation of the laminar or turbulent boundary layer occurs when the shear stress at the wall $\tau_w = 0$, i.e., $\alpha(H) = 0$, $(F_2 = 0)$.

In *laminar* boundary layers with a one-parametric trial solution (3.81), which, together with eqs. (3.86) through (3.92), approximates the Hartree

---

[*]When the dissipation law (3.149) is used together with (3.150), $n - N$ is a function of H. If one uses the new dissipation law by Felsch [27], c_D has to be computed from eq. (3.158), with c_{Di} from eq. (3.156) and c_f from eq. (3.137) in evaluating eq. (4.40).

profiles, the value $\alpha_l = 0$ is associated with $H_{\mathrm{sep}} = 1.515$. In strongly accelerated flows, values of H up to 1.7 are encountered.

For the *turbulent* boundary layer the fact becomes noticeable that in the vicinity of the separation point the turbulent velocity profile really depends on two parameters: besides H, also R_{δ_2} appears as shape parameter. The value $\alpha = 0$ is, strictly speaking, no longer associated with just a single value H_{sep}. One finds approximately $1.57 > H_{\mathrm{sep}} > 1.50$ in the range $10^3 < R_{\delta_2} < 10^5$. The smaller values of H_{sep} correspond to the higher Reynolds numbers. We have approximately

$$(H_{\mathrm{sep}})_{\mathrm{turb}} \approx (H_{\mathrm{sep}})_{\mathrm{lam}} = 1.515 \qquad (4.41)$$

4.3.2 Method of solution

The two ordinary differential equations of first order (4.31) and (4.38) represent a simultaneous system of equations for the two variables $Z(x)$ and $H(x)$. Because of the involved dependence of the universal functions F_1 through F_4 and H^* on the shape parameter H, it is generally impossible to separate the variables.

There are numerous proposals to solve the system of equations (4.31) and (4.38) for incompressible flow, among which the most noted and most frequently applied is the method by Truckenbrodt [138]. He achieves a simplification of the differential equations through linearization and averaging of the universal functions, such that quadratures become possible both for laminar and turbulent boundary layers, similar to the quadratures in Computational Method I described in Sect. 4.2. This principle of solution is, however, not immediately applicable to compressible flow. For turbulent boundary layers, moreover, the likely uncertainties resulting from the forced linearization, in particular, near a separation point, are relatively large (cf., for example, the investigations by Eppler [22, 23].

A thoroughly tried computational method is therefore here presented which applies equally well to both laminar and turbulent boundary layers, in incompressible and compressible flow, without and with heat transfer, and therefore does not require simplifying assumptions on the universal functions. In this method the given velocity distribution $u_\delta(x)$ is again replaced by a piecewise linear function, as in Computational Method I, [eqs. (4.15) et seq.]. The approximation $u_\delta(x)$ can be made arbitrarily precise by choosing the step Δx sufficiently small. On these small intervals, average values may be taken for the shape parameter H, the Reynolds number R_{δ_2}, the Mach number M_δ, and the parameter Θ for heat transfer. This process results in corresponding constant average values for the universal functions F_1 through F_4 in eqs. (4.31) and (4.38).

With this assumption, eqs. (4.31) and (4.38) are immediately integrable in closed form on the interval Δx, as was the case with eq. (4.12) in Computational Method I. We may therefore solve these equations again in step-by-step fashion with any desired degree of precision for each integration interval.

Since the averages of H and Z (or R_{δ_2}) have to be estimated before proceeding to the next interval Δx, in order to compute the values of H and Z at the end of the interval, it is necessary to iterate (which is different from Computational Method I). In general, it suffices to iterate only once.

In analogy to eqs. (4.16) through (4.18), one obtains the following stepping formulas:

For the momentum law

$$\frac{Z_i}{Z_{i-1}} = A_Z + B_Z \overline{F_2} \frac{\Delta x}{Z_{i-1}} , \tag{4.42}$$

For the energy law

$$\frac{H_i^*}{H_{i-1}^*} - 1 = \frac{\Delta H^*}{H_{i-1}^*} = A_H - 1 + B_H \overline{F_4} \frac{2}{Z_i + Z_{i-1}} \cdot \frac{\Delta x}{H_{i-1}^*} , \tag{4.43}$$

where

$$A_Z = \left(\frac{u_{\delta_{i-1}}}{u_{\delta_i}} \right)^{\overline{F_1^*}} ; \tag{4.44}$$

$$B_Z = \frac{1 - A_Z \dfrac{u_{\delta_{i-1}}}{u_{\delta_i}}}{(1 + \overline{F_1^*}) \left(1 - \dfrac{u_{\delta_{i-1}}}{u_{\delta_i}} \right)} ; \tag{4.45}$$

$$A_H = \left(\frac{u_{\delta_{i-1}}}{u_{\delta_i}} \right)^{\overline{F_3}} ; \tag{4.46}$$

$$B_H = \frac{1 - A_H \dfrac{u_{\delta_{i-1}}}{u_{\delta_i}}}{(1 + \overline{F_3}) \left(1 - \dfrac{u_{\delta_{i-1}}}{u_{\delta_i}} \right)} . \tag{4.47}$$

Using eq. (4.35) for $H^*(H, M_\delta, \Theta)$, we find from the analytic representation in Section 4.3.3 also $H(x)$, since $M_\delta(x)$ and $\Theta(x)$ are given with the problem.

For extensive numerical computations it is worthwhile to program the method for an electronic computer. In the case of incompressible flow,

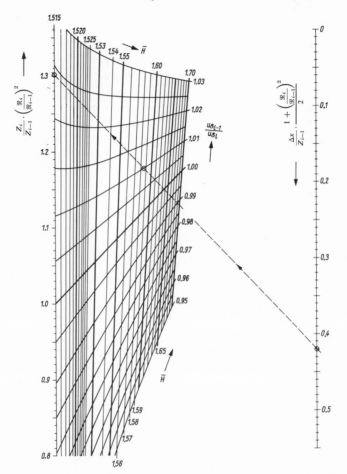

Fig. 4.3. Nomogram for the computation of laminar boundary layers with constant material properties (incompressible flow) based on Method II (see Appendix I.3); determination of $Z(x)$. Given are: $u_{\delta_{i-1}}/u_{\delta_i}$, $\overline{H}$, and $\Delta x/Z_{i-1}$. For axisymmetric boundary layers the cross-sectional radius $\Re(x)$ is known and thus also $\Re_{i-1}/\Re_i$, $\overline{H} = (H_i + H_{i-1})/2$, $\Delta H = H_i - H_{i-1}$.

the computations are substantially facilitated by the use of the nomograms in Figs. 4.3 through 4.7. The Computational Scheme A (p. 269) shows how to proceed best in the computation when using the analytic representation of the universal functions (see Sect. 4.3.3 and Appendix I.3).

4.3.3 Universal functions for Computational Method II

As a consequence of the definitions for the integral quantities δ_1, δ_2, δ_3, and δ_4 [eqs. (3.23), (3.24), (3.44), and (3.45)] and of the coupling

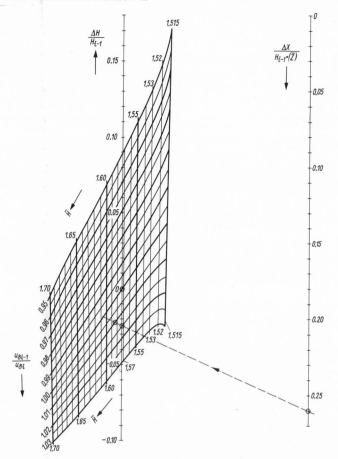

Fig. 4.4. Nomogram for the computation of laminar boundary layers with
constant material properties (incompressible flow) based on Method II
(Appendix I.3). Determination of $H(x)$. Given are: $u_{\delta_{i-1}}/u_{\delta_i}$, $\overline{H}$, and $\Delta x/$
$(H_{i-1} \cdot \overline{Z})$. For axisymmetric boundary layers the cross-sectional radius $\mathfrak{N}(x)$
is given and thus also $\mathfrak{N}_{i-1}/\mathfrak{N}_i$.

$$\overline{H} = (H_i + H_{i-1})/2; \quad \Delta H = H_i - H_{i-1}; \quad \overline{Z} = (Z_i + Z_{i-1})/2.$$

law (2.156), we find general connections that are independent of the special
trial solution for u/u_δ and are thus equally valid for laminar and turbulent
boundary layers. Based on these relations, it is then relatively easy to
represent the universal functions analytically, such that the three quanti-
ties H, M_δ, and Θ appear explicitly. This makes it also possible to develop
the functions for the compressible case (without and with heat transfer)
easily from the functions for the case with constant material properties
(incompressible case).

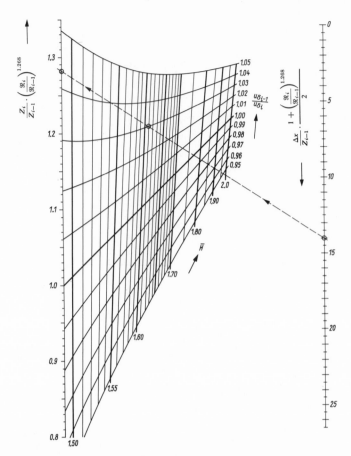

Fig. 4.5. Nomogram for the computation of turbulent boundary layers with constant material properties (incompressible flow) based on Method II (Appendix I.3); determination of $Z(x)$. Given are: $u_{\delta_{i-1}}/u_{\delta_i}$, $\overline{H}$, and $\Delta x/Z_{i-1}$.

$$\overline{H} = (H_i + H_{i-1})/2.$$

We start out by transforming the ratio δ_1/δ_2, which appears in the universal functions F_1 and F_3. A trivial manipulation yields for the displacement thickness δ_1

$$\delta_1 = \int_0^\delta \left(1 - \frac{\rho u}{\rho_\delta u_\delta}\right) dy = \int_0^\delta \left(1 - \frac{u}{u_\delta}\right) dy + \int_0^\delta \frac{\rho u}{\rho_\delta u_\delta} \left(\frac{\rho_\delta}{\rho} - 1\right) dy$$

$$= (\delta_1)_u + \delta_4. \tag{4.48}$$

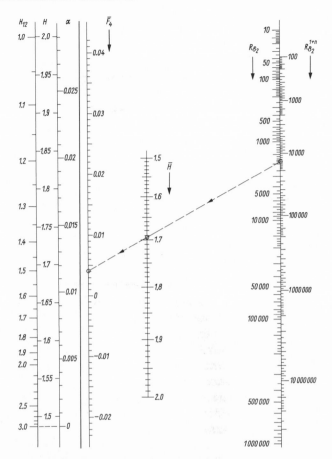

Fig. 4.6. Nomogram for the computation of turbulent boundary layers with constant material properties (incompressible flow) based on Method II (see Appendix I.3); determination of $\overline{F}_4$. Given are $\overline{H}$ and R_{δ_2}.

$$\overline{H} = (H_i + H_{i-1})/2; \quad \overline{F}_4 = (F_{4_i} + F_{4_{i-1}})/2.$$

For the "density-loss thickness" δ_4 we find, with the help of eqs. (2.156) and (2.161)

$$\delta_4 = \int_0^\delta \frac{\rho u}{\rho_\delta u_\delta} \left(\frac{T}{T_\delta} - 1 \right) dy = r \frac{\kappa - 1}{2} \mathrm{M}_\delta^2 \int_0^\delta \frac{\rho u}{\rho_\delta u_\delta} \left[1 - \left(\frac{u}{u_\delta} \right)^2 \right] dy$$

$$- r \frac{\kappa - 1}{2} \mathrm{M}_\delta^2 \Theta \int_0^\delta \frac{\rho u}{\rho_\delta u_\delta} \left(1 - \frac{u}{u_\delta} \right) dy = r \frac{\kappa - 1}{2} \mathrm{M}_\delta^2 (\delta_3 - \Theta \delta_2).$$

$$(4.49)$$

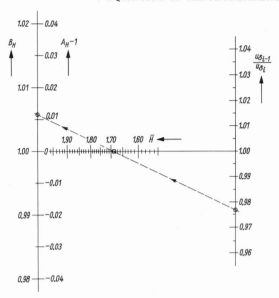

Fig. 4.7. Nomogram for the computation of turbulent boundary layers with constant material properties (incompressible flow) based on Method II (see Appendix I.3); determination of $H(x)$. Given are: $u_{\delta_{i-1}}/u_{\delta_i}$ and $\overline{H}$.

$$\overline{H} = (H_i + H_{i-1})/2;$$
$$\Delta H = H_i - H_{i-1};$$
$$\frac{\Delta H}{H_{i-1}} = A_{\mathrm{H}} - 1 + B_{\mathrm{H}} \cdot \frac{\Delta x \cdot \overline{F_4}}{H_{i-1} \cdot \overline{Z}}.$$

This results in the ratio

$$\frac{\delta_4}{\delta_2} = r \frac{\kappa - 1}{2} \mathrm{M}_\delta^2 \left(\frac{\delta_3}{\delta_2} - \Theta \right) = r \frac{\kappa - 1}{2} \mathrm{M}_\delta^2 (H^* - \Theta), \qquad (4.50)$$

and from eq. (4.48) there follows, with eq. (4.49),

$$\frac{\delta_1}{\delta_2} = \frac{(\delta_1)_\mathrm{u}}{\delta_2} + \frac{\delta_4}{\delta_2} = \left(\frac{\delta_1}{\delta_2} \right)_\mathrm{u} \frac{(\delta_2)_\mathrm{u}}{\delta_2} + r \frac{\kappa - 1}{2} \mathrm{M}_\delta^2 (H^* - \Theta). \qquad (4.51)$$

We now need analytical expressions for $\delta_2/(\delta_2)_\mathrm{u}$, eq. (3.130), and $H^*(H, \mathrm{M}_\delta, \Theta)$.

Equation (3.130) reads:

$$\frac{\delta_2}{(\delta_2)_\mathrm{u}} = \frac{\displaystyle\int_0^\delta \frac{\rho u}{\rho_\delta u_\delta} \left(1 - \frac{u}{u_\delta} \right) \mathrm{d}y}{\displaystyle\int_0^\delta \frac{u}{u_\delta} \left(1 - \frac{u}{u_\delta} \right) \mathrm{d}y}. \qquad (4.52)$$

This expression may be written as

$$\frac{\delta_2}{(\delta_2)_u} = \frac{\bar{\rho}}{\rho_\delta} = \frac{T_\delta}{\bar{T}} \tag{4.53}$$

if it is possible to pull a suitable average for the density (or temperature) ratio $\bar{\rho}/\rho_\delta = T_\delta/\bar{T}$ outside the integral. With T/T_δ from eq. (2.156), it follows from eq. (4.53) that

$$\frac{\delta_2}{(\delta_2)_u} = \frac{1}{1 + r\dfrac{\kappa - 1}{2}\,\mathrm{M}_\delta^2\left[\left(1 + \dfrac{\bar{u}}{u_\delta}\right) - \Theta\right]\left(1 - \dfrac{\bar{u}}{u_\delta}\right)}, \tag{4.54}$$

where the average velocity $\bar{u}/u_\delta$, as yet unknown, is suitably determined via the definitions of δ_3 and δ_2 as follows:

$$\delta_3 = \int_0^\delta \frac{\rho u}{\rho_\delta u_\delta}\left[1 - \left(\frac{u}{u_\delta}\right)^2\right]\mathrm{d}y = \left(1 + \frac{\bar{u}}{u_\delta}\right)\delta_2, \tag{4.55}$$

and thus

$$1 + \frac{\bar{u}}{u_\delta} = H^*. \tag{4.55a}$$

Substitution of this expression into (4.54) yields the final expression:

$$\frac{\delta_2}{(\delta_2)_u} = \frac{1}{1 + r\dfrac{\kappa - 1}{2}\,\mathrm{M}_\delta^2(H^* - \Theta)(2 - H^*)}. \tag{4.56}$$

We suspect, because of the definition in eq. (4.35), that the function $H^*(H, \mathrm{M}_\delta, \Theta)$ satisfies in a first approximation

$$H^* \approx H \tag{4.57}$$

(because of the factor ρ/ρ_δ, which appears in the integrands of δ_2 and δ_3). It is therefore suggested to write

$$H^* = H \cdot \psi(H, \mathrm{M}_\delta, \Theta), \tag{4.58}$$

where ψ represents a correction function. Based on certain limiting cases, for which ψ can be estimated, Jischa [53] established the relation

$$\psi(H, \mathrm{M}_\delta, \Theta) = 1 + \frac{(\psi_{12} - 1)\,\mathrm{M}_\delta}{\mathrm{M}_\delta + \dfrac{\psi_{12} - 1}{\psi_0'}}. \tag{4.59}$$

Here the following abbreviations were used:

$$\psi_{12}(H, \Theta) = \frac{2 - (\delta_1)_u/\delta}{H} \Theta + \frac{1 - (\delta_1)_u/\delta}{H \cdot g} (1 - \Theta);$$

$$g(H) = \int_0^\delta \frac{u/u_\delta}{1 + u/u_\delta} \, dy; \qquad\qquad (4.60)$$

$$\psi_0'(H, \Theta) \approx 0.0114(2 - H)(2 - \Theta)^{0.8}.$$

Appendix I.3 contains analytic expressions for $g(H)$ and $(\delta_1)_u/\delta = f(H)$, both for laminar and turbulent boundary layers. The correction function ψ is between $\psi = 1$ for $M_\delta = 0$ and $\psi \approx 1.18$ for $M_\delta \gg 1$, the latter value being dependent on the heat transfer parameter Θ.

There remains only the question of the analytic representation of the quantity β, which occurs in the function F_4, eq. (4.40). For $\beta(H, M_\delta, \Theta)$ the representation

$$\beta(H, M_\delta, \Theta) = \beta_u(H) \cdot \chi(M_\delta, \Theta) \qquad\qquad (4.61)$$

is possible for both laminar and turbulent boundary layers. For turbulent boundary layer we found already from eq. (3.143) that $\chi \equiv 1$. For laminar boundary layer, we find for β, based on eq. (3.141) and eq. (1.7) that

$$\beta_l = \int_0^{\delta/(\delta_2)u} \frac{\mu}{\mu_w} \left[\frac{\partial(u/u_\delta)}{\partial(y/(\delta_2)_u)} \right]^2 d[y/(\delta_2)_u] = \beta_l(H, M_\delta, \Theta). \qquad (4.62)$$

In order to simplify the evaluation of eq. (4.62) we determine an average value based on the approximations (1.8) and (1.9)

$$\frac{\bar{\mu}}{\mu_w} = \left(\frac{\bar{T}}{T_w}\right)^\omega = \left(\frac{\bar{T}}{T_\delta}\right)^\omega \left(\frac{T_\delta}{T_w}\right)^\omega = \left(\frac{\bar{T}}{T_\delta}\right)^\omega \left(\frac{1}{a}\right)^\omega \qquad (4.63)$$

[a from eq. (2.157)] and pull this average value out from under integral (4.62). We may then represent β_l in the form

$$\beta_l = \beta_{u_l}(H) \cdot \chi(M_\delta, \Theta, \omega) \qquad\qquad (4.64)$$

[ω from eq. (1.9)], where

$$\beta_{u_l}(H) = \int_0^{\delta/(\delta_2)u} \left[\frac{\partial(u/u_\delta)}{\partial(u/(\delta_2)_u)} \right]^2 d[y/(\delta_2)_u], \qquad\qquad (4.65)$$

and [because of eqs. (2.157) through (2.159)]

$$\chi(M_\delta, \Theta, \omega) = \left(\frac{1}{a}\right)^\omega \left(\frac{T}{T_\delta}\right)^\omega = \left(\frac{1}{a}\right)^\omega \left[\int_0^1 \left\{a + b\frac{u}{u_\delta} + c\left(\frac{u}{u_\delta}\right)^2\right\} d\frac{u}{(u_\delta)}\right]^\omega$$

$$= \left(\frac{1}{a}\right)^\omega \left(1 + \frac{2}{3} r \frac{\kappa - 1}{2} M_\delta^2 - \frac{b}{2}\right)^\omega$$

$$= \left(\frac{1}{a}\right)^\omega \left[1 + \frac{2}{3} r \frac{\kappa - 1}{2} M_\delta^2 \left(1 - \frac{3}{4}\Theta\right)\right]^\omega. \qquad (4.66)$$

A comparison of eq. (4.66) with a more accurate numerical evaluation of β_l in eq. (4.62), based on the velocity profiles of Hartree, suggests a modification of eq. (4.66) such that χ still depends slightly upon H:

$$\chi(M_\delta, \Theta, \omega, H) = \left(\frac{1}{a}\right)^\omega \left[1 + k_1(H) r \frac{\kappa - 1}{2} M_\delta^2 - k_2(H) \cdot b\right]^\omega,$$
$$\qquad (4.67)$$

$$k_1(H) = 1.160H - 1.072, \qquad (4.68)$$

$$k_2(H) = 2H - 2.581. \qquad (4.69)$$

For Mach numbers $M_\delta < 5$, the difference between (4.66) and (4.67) amounts to only a few percent.

All the universal functions occurring in Computational Method II have thus been reduced to the H-dependent functions characterizing the case with constant material properties (incompressible), and two "correction functions," ψ and χ. These universal functions and the correction functions can be interpolated by the analytic expressions found in Appendix I.3.

4.4 Computational Method III (with Momentum Law, Energy Law, and Compatibility Condition, for Laminar Boundary Layer only)

Computational Method III is based on eqs. (3.28), (3.46), and (3.62), a two-parametric trial solution, (3.97) through (3.99) for the laminar velocity profile, and the expressions (2.156), and (2.161) for the temperature and density profiles. In order to account for compressibility and heat transfer, we follow Geropp [33] in accepting a suggestion by Howarth [48] to begin by transforming the system of Prandtl's boundary layer eqs. (2.20), (2.21). By this device, the solution of the simultaneous system of three equations in this method is simplified for the case of compressible flow with heat transfer. We put

$$X = x, \tag{4.70}$$

$$Y = \int_0^y \left(\frac{\rho}{\rho_\infty} \frac{T_\infty}{T}\right)^{1/2} dy. \tag{4.71}$$

Using the equation of state for ideal gases

$$\frac{\rho}{\rho_\infty} = \frac{p}{p_\infty} \frac{T_\infty}{T}, \tag{4.72}$$

and Prandtl's assumption $\partial p/\partial y = 0$, we may rewrite eq. (4.71) as follows:

$$Y = \left(\frac{p}{p_\infty}\right)^{1/2} \int_0^y \frac{T_\infty}{T} dy. \tag{4.73}$$

Geropp [33] shows that the restricting assumption $\mu \sim T$ is unnecessary in the original transformation by Howarth. Instead, it is possible to choose an arbitrary viscosity law $\mu(T)$, such as for example the very precise empirical law (1.7). In order to obtain the dimensionless quantity η for eq. (3.97) it is necessary to normalize the transformed distance from the wall, Y, by dividing it by the transformed boundary layer thickness Δ:

$$\eta = Y/\Delta. \tag{4.74}$$

From the transformed boundary layer quantities

$$\Delta_1 = \int_0^\Delta \left(1 - \frac{u}{u_\delta}\right) dY = (\delta_1 - \delta) \frac{T_\infty}{T_\delta} \sqrt{\frac{p}{p_\infty}} + \Delta, \tag{4.75}$$

$$\Delta_2 = \int_0^\Delta \frac{u}{u_\delta} \left(1 - \frac{u}{u_\delta}\right) dY = \delta_2 \frac{T_\infty}{T_\delta} \sqrt{\frac{p}{p_\infty}}, \tag{4.76}$$

$$\Delta_3 = \int_0^\Delta \frac{u}{u_\delta} \left[1 - \left(\frac{u}{u_\delta}\right)^2\right] dY = \delta_3 \frac{T_\infty}{T_\delta} \sqrt{\frac{p}{p_\infty}}, \tag{4.77}$$

$$Z_J = \Delta_2 \frac{\rho_\infty u_\delta \Delta_2}{\mu_\infty}, \tag{4.78}$$

$$Z_E = \Delta_3 \frac{\rho_\infty u_\delta \Delta_2}{\mu_\infty}, \tag{4.79}$$

one obtains the momentum and energy laws in the form

$$Z_J' + Z_J G_J - F_J^* = 0, \tag{4.80}$$

$$Z_E' + Z_E G_E - F_E^* = 0. \tag{4.81}$$

The new quantities occurring in these equations are defined as

$$G_J = \frac{u'_\delta}{u_\delta}\left\{4 + M_\delta^2\left[r\frac{\kappa-1}{2}\left(3 - \frac{2}{r} - \Theta\right) - 1\right]\right\},$$ (4.82)

$$F_J^* = 2\frac{\mu_w}{\mu_\infty}\frac{T_\infty}{T_w}\left[\frac{\partial(u/u_\delta)}{\partial(Y/\Delta_2)}\right]_w - \left(2\frac{\Delta_1}{\Delta_2} - 1\right)\frac{T_w}{T_\delta}Z_J\frac{u'_\delta}{u_\delta},$$ (4.83)

$$G_E = \frac{u'_\delta}{u_\delta}\left\{4 + M_\delta^2\left[r\frac{\kappa-1}{2}\left(3 - \frac{2}{r} - 2\Theta\frac{\Delta_2}{\Delta_3}\right) - 1\right]\right\},$$ (4.84)

$$F_E^* = 2\int_0^{\Delta/\Delta_2}\frac{\mu}{\mu_\infty}\cdot\frac{T_\infty}{T}\left[\frac{\partial(u/u_\delta)}{\partial(Y/\Delta_2)}\right]^2 \mathrm{d}(Y/\Delta_2)$$

$$+ \frac{\Delta_3}{\Delta_2}\left\{\frac{\mu_w}{\mu_\infty}\cdot\frac{T_\infty}{T_w}\left[\frac{\partial(u/u_\delta)}{\partial(Y/\Delta_2)}\right]_w - \frac{\Delta_1}{\Delta_2}\frac{T_w}{T_\delta}Z_J\frac{u'_\delta}{u_\delta}\right\}.$$ (4.85)

G_J is known from the problem data, as is G_E, and they are precisely known when $\Theta = 0$ (for insulated wall) and they are reasonably well known with $(\Delta_3/\Delta_2)_{\text{average}} \approx 1.6 = \text{const.}$, even for $\Theta \neq 0$ (Δ_3/Δ_2 assumes values between 1.5 and 1.7).

In terms of the transformed quantities, the compatibility condition (3.62) may be written as

$$\left[\frac{\partial\left(\mu\frac{T_\infty}{T}\frac{\partial u}{\partial Y}\right)}{\partial Y}\right]_w = -\rho_\infty u_\delta u'_\delta\frac{T_w}{T_\delta}.$$ (4.86)

In order to characterize the laminar velocity profile, two shape parameters are introduced (which become the two unknowns of the theory)

$$\frac{\Delta_3}{\Delta_2} = \frac{\delta_3}{\delta_2} = H^*,$$ (4.87)

$$\Gamma^* = -\left[\frac{\partial^2(u/u_\delta)}{\partial(Y/\Delta_2)^2}\right]_w$$ (4.88)

[see also eqs. (4.36) and (4.6)]. A third unknown of the theory is the quantity Z_J in eq. (4.78), which is a measure for the thickness of the boundary layer. Because of eq. (4.86) the following relations holds

$$\Gamma^* = Z_J\frac{u'_\delta}{u_\delta}\cdot\frac{T_w}{T_\delta},$$ (4.89)

and serves along with (4.80) and (4.81) as the third equation in the theory.

It should be noted that the quantities H^* and Γ^* are pure shape parameters only in the transformed coordinate system X, Y. In an x, y system, on the other hand, these quantities are functions of H, M_δ, and Θ [for definitions of H and Θ, see eqs. (4.36) and (2.160)].

A closed-form solution of the simultaneous system of three equations (4.80), (4.81), (4.89) with the unknowns H^*, Γ^*, and Z_J is not possible in the general case. However, a step-by-step integration of the equations based on the same principles as those in Computational Method II turned out to be practical for this theory as well. Integration over an interval $\Delta x = x_i - x_{i-1}$ yields the following stepping formulas

$$\frac{(Z_J)_i}{(Z_J)_{i-1}} = A_J + B_J \cdot \frac{\overline{F}_J^*}{(Z_J)_{i-1}}, \tag{4.90}$$

$$\frac{(Z_E)_i}{(Z_E)_{i-1}} = A_E + B_E \cdot \frac{\overline{F}_E^*}{(Z_E)_{i-1}}. \tag{4.91}$$

The quantities $\overline{F}_J^*$ and $\overline{F}_E^*$ are averages of F_J^* and F_E^* in the integration interval, and A_J, B_J, and A_E, and B_E are known with the problem data (see Computational Scheme B, p. 276). The values of the shape parameters H^* and Γ^* at the end point x_i of an interval are then obtained from

$$\frac{(Z_E)_i}{(Z_J)_i} = H^*, \tag{4.92}$$

$$\Gamma_i^* = (Z_J)_i \left(\frac{u_\delta'}{u_\delta} \cdot \frac{T_w}{T_\delta}\right)_i. \tag{4.93}$$

The universal functions F_J^* and F_E^* have been computed by Geropp [33] for two-parametric velocity profiles of the type (3.97) with (3.98) and (3.99) and also graphically displayed (see Figs. 4.8, 4.9, and 4.10). It turns out again that these functions, as in Computational Method II, can be composed of two parts, one that depends only on the two shape parameters H^* and Γ^*, while the other terms express the influence of the Mach number and of the heat transfer in analytic form. Equations (4.82) through (4.85) show that the functions G_J, F_J^*, G_E, and F_E^* are greatly simplified for $M_\delta = 0$, and $\Theta = 0$. Practical computation proceeds best according to Computational Scheme B.

In order to obtain the actual velocity profiles that occur in compressible flow (in the physical x, y plane) and the corresponding boundary layer quantities such as δ_1, δ_2, and δ_3, it is necessary to transform back by eqs. (4.70) through (4.77). This results in a distortion of the y coordinate, which modifies the original velocity profiles in expression (3.97). Thus, the coordinate transformation offers an additional choice of shapes for the velocity profiles in the case of the compressible flow. This property of Geropp's theory leads one to expect high accuracy for compressible boundary layer computations (see the examples in Chapter 6).

We mention again that this computational method has been developed so far for laminar boundary layers only.

Figs. 4.8 through 4.10, pp. 152–154. Graphs of the universal functions F_J, F_E and $[\partial(u/u_\delta)/\partial(Y/\Delta_2)]_{Y=0}$ for the two-parameter method by Geropp [33] (Method III, see Appendix I.4).

– – – limiting curve for physically meaningful velocity profiles

– · – locus where a_3 becomes zero. In the region above this curve, $u/u_\delta = 1 - (1 - \eta)^n(1 + a_1\eta + a_2\eta^2)$; and below this curve, $u/u_\delta = 1 - (1 - \eta)^n(1 + a_1\eta + a_2\eta^2 + a_3\eta^3)$;

– ·· – locus for $[\partial(u/u_\delta)/\partial(Y/\Delta_2)]_{Y=0} = 0$.

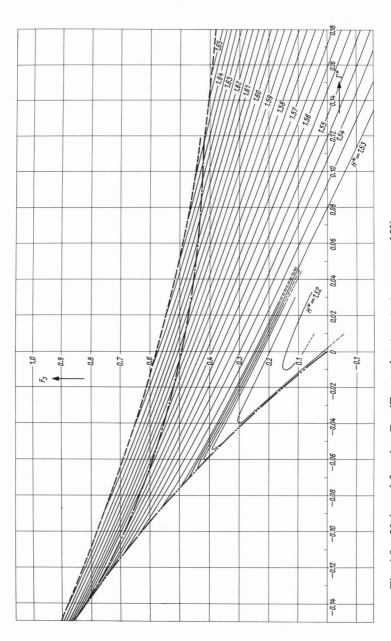

Fig. 4.8. Universal function F_J. (For explanatory text, see page 151).

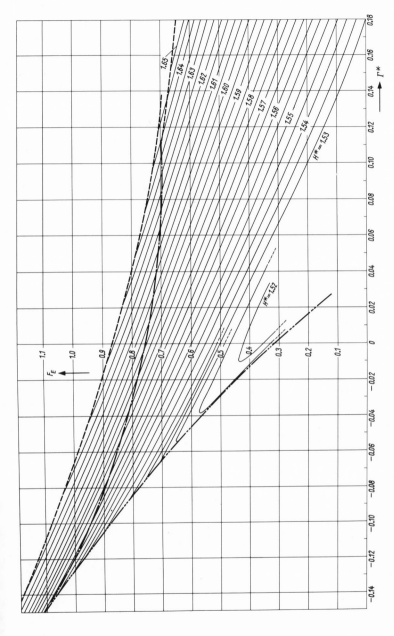

Fig. 4.9. Universal function F_E. (For explanatory text, see page 151).

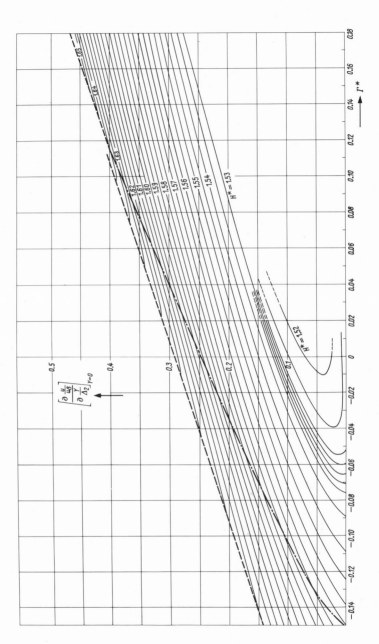

Fig. 4.10.   Universal function $[\partial(u/u_\delta)/\partial(Y/\Delta_2)]_{Y=0}$. (For explanatory text, see page 151).

5 Important Special Problems in Boundary Layer Theory

5.1 Determination of the Transition Point of a Laminar Boundary Layer

The approximation theory described in the preceding chapters was developed under the assumption that the boundary layer is either in the laminar or in the turbulent state. Observations show that every boundary layer starts out being laminar, although often only for a very short distance (at a stagnation point or a pointed leading edge), and then becomes turbulent sooner or later (transition), depending on the given flow conditions, such as the Reynolds number R_x formed with the distance from the leading edge, the pressure gradient, the degree of turbulence of the external flow, and the surface roughness.

Reynolds [106] and Lord Rayleigh [102] as early as 1880 interpreted the emergence of turbulence in the boundary layer to be caused by an instability of the laminar boundary layer, which can be derived from the Navier-Stokes equations. A more detailed mathematical scrutiny of this question centers on the problem of deciding whether a small nonstationary disturbance of the velocity within the laminar boundary layer is amplified or damped. Not until 1929 did Tollmien [136] succeed in overcoming the enormous mathematical difficulties of this problem when he proved the existence of the stability criterion conjectured by Reynolds and Lord Rayleigh, at first only for flows without pressure gradients. The result of Tollmien's theory, obtained numerically, is reproduced in Fig. 5.1 as the connection between values of the Reynolds number $R_{\delta_1} = \rho u_\infty \delta_1/\mu$ and the displacement thickness δ_1 of the boundary layer, normalized by the wave length Λ of the disturbance ($2\pi/\Lambda$ is the wave number of the perturbation, which is assumed to be periodic). The parameter pairs R_{δ_1},

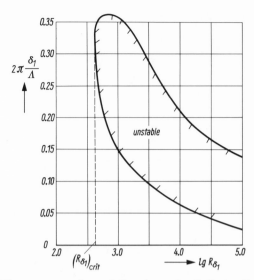

Fig. 5.1.   Indifference curve for stability theory by Tollmien-Schlichting. For points ($2\pi\,\delta_1/\Lambda$, R_{δ_1}) inside this curve, amplification of perturbations is to be expected. Λ = wavelength of perturbation.

$2\pi\delta_1/\Lambda$ inside the "indifference" curve characterize a nonstable state of flow.

Schlichting [112] and Pretsch [98] have augmented this result by analogous computations for the case of one-parametric exact solutions, such as the "similar solutions" (see Section 2.2.1), as examples for flows with pressure gradients (see Fig. 5.2). A complete experimental verification of these results of stability theory was given in 1943 by Schubauer and Skramstad [119]. An explicit presentation of this theory and the numerous experimental investigations concerning the transition problem can be found in Schlichting [111]. Only those results of the theory will be given here that are needed for the application of the approximation theory to boundary layers presented in this book.

It is necessary to remind ourselves of the limitations of stability theory, namely, that so far only the computation of the indifference curve as the locus of the "neutral oscillations" (Figs. 5.1 and 5.2) has been possible, which separates the domains of the damped and amplified oscillations, although it is possible to make a statement about the size of the amplification of a perturbation (Schlichting [113], Pretsch [99], Shen [122]). But any a priori determination of the actual transition point is still outstanding. In flows with steady or intermittent pressure increase, however, it is

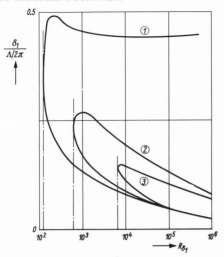

Fig. 5.2. Influence of a pressure rise or drop upon the indifference curve.
These are results obtained by Pretsch [98] for flows of the type $u_\delta \sim x^m$
(pressure rise for $m < 0$, pressure drop for $m > 0$) which lead to one-
parametric velocity profiles of Hartree type (similar solutions); see also
Section 2.2.1.2. The smallest value of the abscissa on these indifference curves
is denoted by $(R_{\delta_1})_{crit}$ (see dot-dash vertical lines).

    ① $m = -0.05$; $H = 1.553$
    ② $m = 0$; $H = 1.572$
    ③ $m = 1$; $H = 1.625$

possible to give bounds for the transition point with high certainty: the
indifference point from the Tollmien-Schlichting theory, characterized by
the critical Reynolds number $(R_{\delta_1})_{crit}$ (the point at the extreme left of
the curves $[(\delta_1)/(\Lambda/2\pi)]$, R_{δ_1} in Figs. 5.1 and 5.2), as the lower bound,
and the laminar separation point (which can be determined from ordinary
boundary layer theory as presented in this book) as the upper bound.

According to Fig. 5.3, the critical Reynolds number decreases rapidly
with decreasing shape parameter H (as the laminar separation point is
approached). At the same time, the amplification rises sharply (qualitative
curve A in Fig. 5.3). From this result of stability theory one may conclude
that the laminar separation point is indeed the latest possible transition
point (at least for sufficiently large Reynolds numbers for which boundary
layer theory holds). In accelerated flow, on the other hand, the increased
value of H results in larger values for $(R_{\delta_1})_{crit}$ and thus transition is
generally prevented. Therefore, in flows where acceleration begins immediately behind the stagnation point, followed by a domain of retarda-

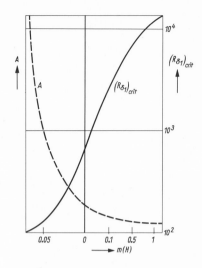

Fig. 5.3. Dependence of $(R_{\delta_1})_{\text{crit}}$ upon the shape parameter of the similar velocity profiles.
—— $(R_{\delta_1})_{\text{crit}}$ plotted against $m(H)$
– – – Qualitative behavior of the amplification factor A as a function of $m(H)$.

tion (as happens, for example, in the flow around a wing profile), the indifference point is found to be immediately behind the point of maximum velocity and not very far from the laminar separation point. The two limiting positions of the transition point are therefore often very close together so that the result of the computation of the turbulent boundary layer is less imprecise than one might suspect from the fundamental uncertainty of the prediction of the transition point.

In incompressible flow, the relation

$$[R_{(\delta_2)u}]_{\text{crit}}(H) = [R_{(\delta_1)u}]_{\text{crit}} \frac{1}{H_{12}} = f(H) \tag{5.1}$$

can be replaced by the following equation as a good approximation

$$\log[R_{(\delta_2)u}]_{\text{crit}} \approx 2.42 + 24.2(H - 1.572). \tag{5.2}$$

It is recommended in any case to compute the turbulent boundary layer for both limiting positions of the transition point. Observation shows that for free-stream flow with low turbulence, for example, in flight in the free atmosphere, and with aerodynamically smooth surfaces, the transition point is closer to the laminar separation point than to the indifference point. It is therefore generally possible to estimate which of the two computations of the turbulent boundary layer is more appropriate for a given example.

It remains to mention that at the transition point the following relations hold:

$$H_{\text{turb}} = H_{\text{lam}} \tag{5.3}$$

$$(\delta_2)_{\text{turb}} = (\delta_2)_{\text{lam}} \tag{5.4}$$

It follows then with eq. (4.29):

$$Z_{\text{turb}} = [R_{\delta_2}^{0.268} \cdot \delta_2]_{\text{turb}} = Z_{\text{lam}}(R_{\delta_2})_{\text{lam}}^{-0.732}$$
$$= [R_{\delta_2} \cdot \delta_2]_{\text{lam}} \cdot (R_{\delta_2})_{\text{lam}}^{-0.732}. \tag{5.5}$$

5.2 Computation of the Boundary Layer along Axisymmetric Contours

In hydrodynamics one often encounters axisymmetric bodies in axial flows (for example, hub shields in jet engines, the fuselage of airplanes, rockets). The boundary layer along such bodies is three-dimensional, though. But for axial symmetry the resulting boundary layer theory is hardly more complicated than in the purely two-dimensional case.

Let x in Figure 5.4 be the coordinate along the generator of a body of revolution, and let y be coordinate perpendicular to x, and $\mathfrak{R} = \mathfrak{R}(x)$ the cross-sectional radius the body of revolution, then continuity and the equilibrium of forces in x direction can be expressed, as in Section 2.1, as follows:

$$\frac{\partial(\rho u \mathfrak{R})}{\partial x} + \frac{\partial(\rho v \mathfrak{R})}{\partial y} = 0, \tag{5.6}$$

$$\rho u \frac{\partial u}{\partial x} + \rho v \frac{\partial u}{\partial y} = -\frac{dp}{dx} + \frac{\partial}{\partial y}\left(\mu \frac{\partial u}{\partial y}\right). \tag{5.7}$$

In contrast to the system (2.20), (2.21), which is valid for two-dimensional flow, only the continuity equation is changed by the appearance of the factor $\mathfrak{R}$. In deriving relations (5.6) and (5.7) we assumed, in addition to

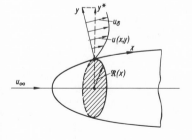

Fig. 5.4.   Illustration of the choice of coordinate system used for axisymmetric boundary layer flow. y^* is defined by eq. (5.14).

Prandtl's boundary layer simplifications (Section 1.7), that the boundary layer thickness δ is small compared to the cross-sectional radius $\mathfrak{R}$ ($\delta/\mathfrak{R} \ll 1$). If we derive integral conditions from (5.6) and (5.7), based on the mathematical principle described in Section 3.2.2, then the momentum law assumes the form

$$Z' + Z\frac{u_\delta'}{u_\delta}\left[F_1 + n\frac{\mu_w'/\mu_w}{u_\delta'/u_\delta} + (1+n)\frac{\mathfrak{R}'/\mathfrak{R}}{u_\delta'/u_\delta}\right] - F_2 = 0, \qquad (5.8)$$

where Z, defined in eq. (4.29), is the parameter for the boundary layer thickness. Compared to eq. (4.31) for the plane case, only the term with the factor $\mathfrak{R}'/\mathfrak{R}$ was added. The energy law (4.38) remains valid without change for axisymmetric flow as well.

Relation (4.42) for the step-by-step computation of $Z(x)$, under the assumption of piecewise linear $u_\delta(x)$, becomes for the axisymmetric case*

$$\frac{Z_i}{Z_{i-1}} = \left(\frac{\mathfrak{R}_{i-1}}{\mathfrak{R}_i}\right)^{1+n}\left[A_Z + B_Z\overline{F_2}\frac{\Delta x}{Z_{i-1}}\frac{1+(\mathfrak{R}_i/\mathfrak{R}_{i-1})^{1+n}}{2}\right]. \qquad (5.9)$$

The functions A_Z and B_Z are determined from eqs. (4.44) and (4.45). For laminar boundary layer we have $n = 1$, for turbulent boundary layer $n = 0.268$.

The single step formula (4.43) together with (4.46) and (4.47) for the quantity $H^*(x)$ remains unchanged, as does the energy law (4.38), even for the axisymmetric case (see also Walz [143]). The nomograms of Figs. 4.3 through 4.7 remain immediately applicable for incompressible flows in the case of axisymmetric boundary layers as well.

We should also mention here that the simple connection between two-dimensional and axisymmetric boundary layers can also be exhibited by a coordinate transformation. Mangler [70] has dealt extensively with this problem. Denoting the coordinates of the axisymmetric case by x, y and the coordinates of the associated two-dimensional case by $\bar{x}$, $\bar{y}$, we have

$$\bar{x} = \int_0^x \left(\frac{\mathfrak{R}(x)}{L}\right)^{1+n} dx, \qquad (5.10)$$

$$\bar{y} = \frac{\mathfrak{R}(x)}{L}y \qquad \text{i.e., } \bar{\delta}_1 = \frac{\mathfrak{R}(x)}{L}\delta_1, \text{ etc.} \qquad (5.11)$$

As in eq. (5.9), n is the exponent of the Reynolds number R_{δ_2} in the shear stress law (3.132) at the wall. Equations (5.10), (5.11) will be used with

*For the derivation of this relation, the cross-sectional radius was assumed to be a linear function of x within each interval Δx.

advantage in the computation of the coefficient of friction of an axisymmetric cone in supersonic flow (see Section 6.2.6).

It should be noted here that the theory of axisymmetric boundary layers remains relatively simple even without the limiting assumption $\delta/\Re \ll 1$. This theory is of interest, for example, in mechanical process control of the axial flows along very slim bodies, such as threads produced by fiberization in pressure jets. When formulating the equation of continuity and the equation of motion in x direction according to the principle described in Section 2.1, we obtain, according to Mayer [71], for arbitrarily large values of $y/\Re$ and $\delta/\Re$

$$\frac{\partial}{\partial x}[\rho u(\Re + y^*)] + \frac{\partial}{\partial y^*}[\rho v(\Re + y^*)] = 0, \qquad \text{(continuity)} \quad (5.12)$$

$$\rho u \frac{\partial u}{\partial x} + \rho v \frac{\partial u}{\partial y^*} = -\frac{dp}{dx}$$

$$+ \frac{1}{\Re + y^*} \frac{\partial}{\partial y^*}[\tau(\Re + y^*)], \qquad \text{(momentum law)} \quad (5.13)$$

where

$$y^* = y\sqrt{1 - (d\Re/dx)^2}. \qquad (5.14)$$

Mayer derived the integral conditions for momentum and energy from an approximation theory resulting from (5.12) and (5.13) and tabulated the occurring universal functions for laminar and turbulent boundary layer for incompressible and compressible flow. Numerical examples based on this approximation theory which might shed some light on the fiberization process for thermoplastic materials are discussed extensively by Mayer.

5.3 Computation of Heat Transfer in Strong Temperature and Pressure Gradients in the Direction of Flow

5.3.1 The problem

The rather simple coupling law (2.156) between flow and temperature boundary layers was derived under the assumption that there are no gradients of the wall temperature or of pressure in the direction of flow (x direction). In addition to the assumption $dp/dx = 0$, the temperature at the wall (which can be arbitrarily chosen) is taken $T_W = $ const., that is $dT_W/dx = 0$. Moreover, the Prandtl number is assumed to be $\text{Pr} = 1$ or deviating only slightly from unity.

We pointed out in Section 2.3.6 that, according to past investigations and conclusions [66], a pressure gradient dp/dx has relatively little influence on the connection between flow and temperature boundary layers, but that on the other hand (see for example the investigations of Chapman and Rubesin [7], Schlichting [114] and Dienemann [11]) a temperature gradient dT_W/dx may under certain circumstances have a strong influence on this connection.

For the practical computation of flow and temperature boundary layers, most of all for the computation of the heat transfer, it is now important to know more precisely the limits of validity of the coupling relation (2.156) in order to be able, if necessary, to compute with a more precise relation. In addition to the influences of dT_W/dx and dp/dx, the influence of the Prandtl number Pr should also be accounted for; the Prandtl number for fluids such as water and oil is much larger than unity, but for liquid metals, such as mercury, with Pr ≈ 0.02 at 20° C, it is much less than unity.

Because of the slim chances for finding an exact solution of the general problem as posed by eqs. (2.121), (2.122) (see the exact solutions by Li and Nagamatsu [66] and Geropp [34] for the special case of similar solutions for laminar compressible boundary layers, but with $T_W = $ const.), it is desirable to attempt an approximate solution with integral conditions or compatibility conditions (or with both types of equations simultaneously).

To this end, it is necessary to introduce a trial solution for the temperature profile $T(x, y)/T_\delta(x)$ which is more generally valid than eq. (2.156). This trial solution has to be such that for the case $T_W = T_e$ ($b = 0, \Theta = 0$) the coefficient of the linear term in eq. (2.156) is not automatically zero as is the case with the definition (2.158). The outer boundary condition (2.124), $u/u_\delta = 1, T/T_\delta = 1$, which follows from Crocco's transformation (2.118) should, however, still be satisfied.

5.3.2 Approach for an approximate solution (modified Reynolds analogy)

When the trial solutions are explicitly given, T/T_δ is expanded in powers of the variable y (the distance from the wall) with coefficients which depend upon x (Morris and Smith [75]) or in powers of the variable x with coefficient functions which are determined from ordinary differential equations (Schlichting [114]). In the latter case, the number of terms that have to be used in the power series expansion for the solution is determined by the data of the problem [for example, a given function for the wall temperature $T_W(x)$].

In general we may state that a trial solution is the better, the more properties of exact solutions it encompasses. Among the exact solutions considered in Section 2.3 the solution (2.137) by van Driest [17] is of most general character: it includes the case of arbitrary (constant) Prandtl number Pr for compressible laminar and turbulent boundary layers. Only the influence of dT_W/dx and dp/dx is neglected there.

A temperature gradient dT_W/dx, according to [7] and [114] leads, among others, to the consequence that for vanishing temperature difference between wall and external flow, that is, at the point where $b = (T_e - T_W)/T_\delta = 0$, the heat flow q need not vanish. A generalization of eq. (2.137) in the simplest case may consist of adding an additional term $K(x)$ to the coefficient $(T_e - T_W)/T_\delta$ of the function f_1 in eq. (2.137). The function $K(x)$ is a correction parameter, intended to describe the influence of dT_W/dx and dp/dx as well.

In order to satisfy the boundary conditions

$$u/u_\delta = 1 \ (f_1 = f_2 = 1), \quad T/T_\delta = 1$$

of the exact differential equation with such a modified trial solution, after carrying out the Crocco transformation [10], the coefficient $c = -r(\kappa - 1) M_\delta^2/2$ in eq. (2.137) has to be replaced by $c - K(x)$. A trial solution for gases, constructed according to these considerations, is then given by

$$\frac{T}{T_\delta} = a + [b + K(x)]f_1 + [c - K(x)]f_2 = \frac{\rho_\delta}{\rho}, \tag{5.15}$$

where $i = c_p T$. $K(x)$ is the new third unknown of the approximation theory [besides Z from eq. (4.29) and H from eq. (4.36)]. It is therefore necessary for the determination of $K(x)$ to have an equation in addition to (4.31) and (4.38). The integral condition for the total energy h, eq. (3.57), appears particularly well suited for this purpose. That equation is valid for laminar and turbulent boundary layers for compressible flow and, what appears essential, also for arbitrary (constant) Prandtl number.

5.3.3 Solution for Pr = 1

At the present time a solution has only been developed for the case Pr = 1. In that case, we have $f_1 = u/u_\delta$, $f_2 = (u/u_\delta)^2$, eq. (5.15) simplifies to

$$\frac{T}{T_\delta} = a + [b + K(x)]\frac{u}{u_\delta} + [c - K(x)]\left(\frac{u}{u_\delta}\right)^2. \tag{5.16}$$

Using the definition (2.67) for the total energy h and the relation $\rho/\rho_\delta = T_\delta/T$, one obtains, after elementary transformation from eqs. (3.57) and

(5.16), an ordinary differential equation of first order for $K(x)$ (′ denotes derivative with respect to x):

$$K' + KF_5 - F_6 = 0. \tag{5.17}$$

The coefficients are:

$$F_5 = \frac{u_\delta'/u_\delta}{H^* - 1}\left[1 + \frac{\delta_1}{\delta_2} + (\kappa - 1)\,\mathrm{M}_\delta^2 + H^*(1 - \kappa\mathrm{M}_\delta^2)\right]$$
$$+ \frac{H^{*\prime} + H^*\,\delta_2'/\delta_2}{H^* - 1}\,, \tag{5.18}$$

$$F_6 = \frac{1}{H^* - 1}\left\{b' - b\frac{u_\delta'}{u_\delta}\left[1 + \frac{\delta_1}{\delta_2} + (\kappa - 1)\,\mathrm{M}_\delta^2\right]\right\}. \tag{5.19}$$

The derivatives δ_2' and $H^{*\prime}$ in eq. (5.18) may be eliminated by using (3.28) and (4.38). The pressure gradient dp/dx is accounted for in F_5 through u_δ', and the temperature gradient dT_W/dx in F_6 through b'. Otherwise only quantities of the flow boundary layer occur in F_5 and F_6. When the material properties ρ and μ are variable, these functions contain, although implicitly, also the correction factor $K(x)$. For the universal functions of Section 4.3.3 we find from the modified trial solution (5.16) the following more general relations:

$$\frac{\delta_4}{\delta_2} = r\frac{\kappa - 1}{2}\,\mathrm{M}_\delta^2\,H^* - b + K\,(H^* - 1)$$
$$= r\frac{\kappa - 1}{2}\,\mathrm{M}_\delta^2\,(H^* - \Theta) + K\,(H^* - 1), \tag{5.20}$$

$$\frac{\delta_1}{\delta_2} = H_{12}\frac{(\delta_2)_u}{\delta_2} + r\frac{\kappa - 1}{2}\,\mathrm{M}_\delta^2\,H^* - b + K\,(H^* - 1)$$
$$= H_{12}\frac{(\delta_2)_u}{\delta_2} + r\frac{\kappa - 1}{2}\,\mathrm{M}_\delta^2\,(H^* - \Theta) + K\,(H^* - 1). \tag{5.21}$$

For all other relations, we have to replace b by $b + K$. The modified relation (2.170) for the heat transfer becomes thus

$$\frac{q_w(x)}{\rho_\delta u_\delta^3} = -\frac{1}{2}\,\mathrm{ST}\cdot(b + K)\frac{c_p T_\delta}{u_\delta^2/2}\,, \tag{5.22}$$

or, for compressible media with $\mathrm{M}_\delta > 0$, more simply

$$\frac{q_w(x)}{\rho_\delta u_\delta^3} = -\frac{r}{2}\,\mathrm{ST}\cdot\Theta\cdot\left(1 + \frac{K}{b}\right). \tag{5.23}$$

The modification developed here is obviously essential when K/b approaches magnitude 1.

It is useful to determine the correction function $K(x)$ iteratively from eq. (5.17): One starts out with $K^{(0)} = 0$ and determines approximations for F_5 and F_6; then the equation is solved:

$$K^{(1)'} + K^{(1)}F_5^{(0)} - F_6^{(0)} = 0. \qquad (5.24)$$

From this follow, with the aid of eq. (5.21), first approximations $F_5^{(1)}$ and $F_6^{(1)}$ and then a second approximation $K^{(2)}$, etc. Present experience indicates that the computations converge rapidly, such that $K^{(1)}(x)$ is generally sufficiently accurate.

The case of constant material properties, which was investigated by Schlichting [114] for the flat plate in parallel flow and locally varying wall temperature $T_W(x)$ ($dT_W(x)/dx \neq 0$), leads to closed-form solutions of eq. (5.17), and the same holds for the general flow type $u_\delta \sim x^m$. With these closed-form solutions the influences of the parameters dT_W/dx and dp/dx can be investigated separately in the important case of constant material properties. Details of these and other more general solutions which include the influence of compressibility and turbulent boundary layers will be discussed in detail in Section 6.12.

The method of solution for arbitrary Prandtl number, starting out from eq. (5.15), corresponds essentially to the approach described here for $Pr = 1$. There is additional numerical effort required in that, instead of u/u_δ, the integrals f_1 and f_2 from eqs. (2.126) through (2.131) have to be evaluated, and the functions F_5 and F_6 are now dependent on Pr.

5.4 Computation of the Potential Flow, Drag, and Maximum Lift of Airfoils

5.4.1 Introductory remarks

The creation of lift and forward pushing forces on profiles (for example airfoils or propeller blades of hydraulic machines) is limited by boundary layer separation, as is the pressure gain in diffusers (Section 6.7). The maximum lift of a profile can therefore basically be determined by boundary layer computations when the potential-theoretical velocity distribution $u_\delta(x)$ is known for different angles of attack α_e of the profile against the direction of the primary flow. The computation of the potential flow about arbitrary profiles is a problem that is solved for incompressible media by known methods of conformal mapping (see for example Betz [3]) or by the so-called singularity methods (see for example Jacob and Riegels [52]).

It is even possible to account for a certain feedback of the boundary layer on the potential flow (essentially the displacement effect) (the methods of Theodorsen and Garrick [133] and Pinkerton [90]), which are ordinarily neglected within the framework of the assumptions of Prandtl's boundary layer theory.

In compressible flow, there are well-known methods from gas dynamics at our disposal for the computation of the frictionless external flow (see, for example, Oswatitsch [85] or Zierep [160]).

Besides the maximum lift, the boundary layer computation also gives the viscous drag and the form drag (cf. Section 5.4.3). And eventually it is easily possible to account for the influence of boundary layer suction by single slots on the maximum lift and drag when using the computational method described in Section 5.5.

For the determination of the potential flow about arbitrary profiles, which is required as a starting point for all the applications of boundary layer computations that have been discussed, there are proven computer programs both for the methods of conformal mapping [133, 90], and the singularity methods, see for example [52]. A description follows here of a particularly simple but very powerful method of conformal mapping which is based on Theodorsen's and Pinkerton's work.

5.4.2 Computation of the velocity of potential flow, $u_\delta(x)$, according to Theodorsen-Pinkerton-Walz, allowing for boundary layer feedback

The method of conformal mapping described below unfortunately has the disadvantage of being applicable to plane flow only, but it produces very precise results for arbitrary profiles since it makes it possible to forego the simplifying approximations that are necessary with the singularity methods. The time required (with or without an electronic computer) does not exceed that required by the singularity methods. Moreover, it is possible to account for the boundary layer feedback on the potential flow in a particularly simple manner. Detailed instructions for practically carrying out the method of conformal mapping which is briefly sketched below was given by Schrenk and Walz [117].

If we denote in Fig. 5.5 the vectors pointing to the profile points by

$$\zeta = \xi + i\eta \tag{5.25}$$

and the vectors to the points on the circle by

$$Z = \hat{x} + i\hat{y} = l\,e^{\psi_0 + i\varphi} \tag{5.26}$$

(l = distance of the trailing edge of the profile from the origin of the coordinate system), the potential flow about the profile is uniquely related

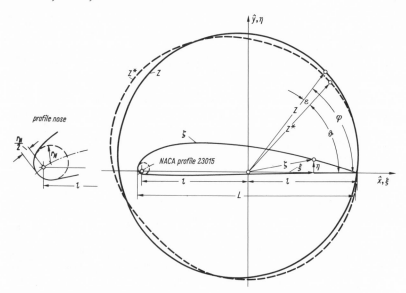

Fig. 5.5.   Illustration of the conformal mapping in two steps according to Theodorsen and Garrick [133].

to the known potential flow about the circle by a conformal mapping

$$\varsigma = F(Z) \tag{5.27}$$

The mapping function F can be determined from the given coordinates ξ, η of the profile. Let $u_{\delta K}(Z)$ be the known velocity distribution at the periphery of the circle; it follows for the velocity distribution $u_\delta(\varsigma)$ along the profile that

$$\left| \frac{u_\delta(\varsigma)}{u_\infty} \right| = \left| \frac{u_{\delta K}(Z)}{u_\infty} \right| \left| \frac{\mathrm{d}Z}{\mathrm{d}\varsigma} \right|. \tag{5.28}$$

According to Theodorsen and Garrick [133], it is useful to proceed in two steps with the mapping. In the first step, the contour of the profile is transformed by the Joukowski mapping function (see [3])

$$\varsigma = Z^* + \frac{l^2}{Z^*} \tag{5.29}$$

into a figure in the Z^* plane which is "similar" to a circle. Should the profile to be mapped be of the special Joukowski type (dashed contour in Fig. 5.5), the problem would have been solved already by this first step of the mapping. The second step of the mapping [analogous to eq. (5.26)] is

$$Z^* = f(Z) = l\, e^{\psi + i\vartheta} = Z \cdot e^{\psi - \psi_0 + i(\vartheta - \varphi)}, \tag{5.30}$$

where

$$\psi_0 = \frac{1}{2\pi} \int_0^{2\pi} \psi(\varphi) \, d\varphi; \tag{5.31}$$

$$l \cdot e^{\psi_0} = R = \text{radius of mapping circle.} \tag{5.32}$$

This step is introduced to correct for the deviation of the given profile from the Joukowski approximation and takes on the character of a correction. We have then

$$\left| \frac{dZ}{d\zeta} \right| = \left| \frac{dZ}{dZ^*} \right| \cdot \left| \frac{dZ^*}{d\zeta} \right|. \tag{5.33}$$

According to Theodorsen and Garrick, eq. (5.30) is solved by iteration. In mapping the points of the figure which is almost a circle (Z^* plane) into the circle itself (Z plane), the difference in the azimuths

$$\varphi - \vartheta = \epsilon \tag{5.34}$$

between the two vectors Z^* and Z, which are correlated to one another, is at first ignored. The logarithm of the deviation

$$\psi(\varphi) = \ln \frac{Z^*}{l} \tag{5.35}$$

is taken to be the real part of the complex mapping function f to which the imaginary part $\epsilon(\varphi)$ is sought. For an arbitrary point φ' the following relation holds

$$\epsilon(\varphi') = \frac{1}{2\pi} \int_0^{2\pi} \psi(\varphi) \operatorname{ctg} \frac{\varphi - \varphi'}{2} \, d\varphi. \tag{5.36}$$

φ, however, is as yet unknown. Since $\psi(\vartheta)$ is given, a first approximation $\epsilon^{(1)}$ with $\varphi^{(0)} = \vartheta$ follows from

$$\epsilon^{(1)}(\varphi^{(0)}) = \epsilon^{(1)}(\vartheta) = \frac{1}{2\pi} \int_0^{2\pi} \psi(\vartheta) \operatorname{ctg} \frac{\vartheta - \vartheta'}{2} \, d\vartheta. \tag{5.37}$$

From $\vartheta + \epsilon^{(1)} = \varphi^{(1)}$ follows a second approximation $\epsilon^{(2)}$, and from this follows $\vartheta + \epsilon^{(2)} = \varphi^{(2)}$. The method converges rather rapidly for reasonably smooth contours. Once $\epsilon(\varphi)$ and $\psi(\varphi)$ are found, there finally follows from eqs. (5.33) to (5.37) for $u_\delta(x)$ the relation (with x the coordinate

measured along the profile contour)

$$\frac{u_\delta(x)}{u_\infty} = \frac{u_{\delta K}}{u_\infty} \left|\frac{dZ}{dZ^*}\right| \left|\frac{dZ^*}{d\zeta}\right| = \frac{u_{\delta K}}{u_\infty} \frac{R}{|Z^*|} \frac{1 + d\epsilon/d\varphi}{\sqrt{1 + \left(\dfrac{d\psi}{d\varphi}\right)^2}} \cdot \left|\frac{Z^{*2}}{Z^{*2} - l^2}\right|,$$

(5.38)

where

$$\frac{u_{\delta K}}{u_\infty} = 2\{\sin(\varphi - \alpha_g) + \sin\alpha_e\}$$

(5.39)

is the velocity distribution along the circle, with circulation included. Here α_e is the effective angle of attack, which is the difference between the geometric angle α_g against the $\hat{x}$ axis of the coordinate system of the mapping, and the zero angle of attack $\epsilon_0 = \epsilon_{\varphi=0} = \epsilon_{\text{trailing edge}}$ (a property of the profile):

$$\alpha_e = \alpha_g - \epsilon_0.$$

(5.40)

For symmetric profiles we have $\epsilon_0 = 0$. Changes in α_g merely influence $u_{\delta K}$ (that is, the circulation about the profile). The mapping function and its derivative $|dZ/d\zeta|$, which occurs in eq. (5.38), are the same for all angles of attack α_g.

The boundary layer feedback on the potential flow, according to Pinkerton [90], consists essentially in an increasingly lower camber of the profile as the effective angle of attack α_e increases, as illustrated in Fig. 5.6, predicated on different displacement thicknesses on the upper and lower parts of the profile. The decrease in camber can be described by a change in the direction of zero lift of the profiles by the angle $\Delta\epsilon_0$. The deviation from the original contour of the profile, which is connected with the decreased camber of the profile, is strongest at the trailing edge and decreases continuously toward the leading edge (because of the lower boundary layer thickness there). Pinkerton describes this decrease of the

Fig. 5.6.　Illustration of the lessening of the camber by a boundary layer with different thicknesses on the pressure and suction sides of an airfoil.
—— original contour of NACA profile 4412.
– – – decambered contours for geometric angles of attack $\alpha_g = 8°$ and $16°$.

deviation from the original shape by modulating the function $\epsilon(\varphi)$ in eq. (5.38) as follows:

$$\epsilon_{\mathrm{m}}(\varphi) = \epsilon(\varphi) + \Delta\epsilon_0 \frac{1 + \cos \varphi}{2}. \tag{5.41}$$

At the leading edge of the profile we have $\varphi \approx 180°$, $\cos \varphi = -1$, and thus $\epsilon_{\mathrm{m}} = \epsilon$. The change of the real part $\psi_{\mathrm{m}}(\varphi)$ based on the expression in eq. (5.41), and the square of its derivative $(d\psi_{\mathrm{m}}/d\varphi)^2$ in eq. (5.38) of the Theodorsen mapping function (5.30) is so small because of the correction character of $\Delta\epsilon_0$ that practically only the derivative of eq. (5.41) enters in eq. (5.38) and we may put $(d\psi_{\mathrm{m}}/d\varphi)^2 \approx (d\psi/d\varphi)^2$.

The mapping function according to Theodorsen, Garrick and Pinkerton is then given with eq. (5.41) by

$$\left|\frac{dZ}{dZ^*}\right|_{\mathrm{m}} = \frac{R}{|Z^*|} \frac{1 + \dfrac{d\epsilon}{d\varphi} - \dfrac{\Delta\epsilon_0}{2}\sin\varphi}{\sqrt{1 + \left(\dfrac{d\psi}{d\varphi}\right)^2}} \tag{5.42}$$

and the explicit formula for the potential flow along the profile:

$$\frac{u_\delta(x)}{u_\infty} = \frac{u_{\delta\mathrm{K}}}{u_\infty} \frac{R}{|Z^*|} \frac{1 + \dfrac{d\epsilon}{d\varphi} - \dfrac{\Delta\epsilon_0}{2}\sin\varphi}{\sqrt{1 + \left(\dfrac{d\psi}{d\varphi}\right)^2}} \cdot \left|\frac{Z^{*2}}{Z^{*2} - l^2}\right|. \tag{5.43}$$

The velocity $u_{\delta\mathrm{K}}$ along the circle is found from eq. (5.39) to

$$\frac{u_{\delta\mathrm{K}}}{u_\infty} = 2\{\sin(\varphi - \alpha_{\mathrm{g}} + \Delta\epsilon_0) + \sin(\alpha_{\mathrm{e}} - \Delta\epsilon_0)\}. \tag{5.44}$$

According to Walz [144] the quantity $\Delta\epsilon_0$, which describes the reduction in camber of the profile, can be represented by an empirical relation which holds with sufficient accuracy for customary profiles in aerodynamics with a thickness ratio d/L of about 9 to 15 percent:

$$\frac{\Delta\epsilon_0}{\alpha_{\mathrm{e}}} \approx 0.2\,[1 - 0.026(16° - \beta_{\mathrm{H}})]. \tag{5.45}$$

Here the quantity β_{H} is the angle, in degrees, of the profile at the trailing edge (i.e., it equals the angle between the two tangents at the profile contour at the trailing edge). The lift coefficient

$$c_{\mathrm{a}} = \frac{A}{\dfrac{\rho_\infty}{2} u_\infty^2 F} \tag{5.46}$$

(A = lift, F = area of the airfoil, L = length of profile) may then be written as

$$c_{\mathrm{a}} = \frac{4\pi R}{L} \sin(\alpha_{\mathrm{e}} - \Delta\epsilon_0) \tag{5.47}$$

where R is given by (5.32).

The computational method described in this section is, in an extended form [145], also applicable to profiles with flaps and the attached dead-water region (wake).

5.4.3 Computation of the total drag of a profile

The local coefficient of drag $c_{\mathrm{f}} = 2\tau_{\mathrm{w}}/\rho_\delta u_\delta^2$ and the total drag coefficient*

$$C_{\mathrm{F}} = \frac{\displaystyle\int_0^1 \tau_{\mathrm{w}}\, \mathrm{d}\left(\frac{x}{L}\right)}{\dfrac{\rho_\infty}{2} u_\infty^2} = \int_0^1 \frac{\rho_\delta}{\rho_\infty}\left(\frac{u_\delta}{u_\infty}\right)^2 c_{\mathrm{f}}\, \mathrm{d}\left(\frac{x}{L}\right) \tag{5.48}$$

(x = direction of the free-stream flow u_∞) of a profile are known after the boundary layer computation with the methods described in Chapter 4 is carried out.

When Prandtl's assumption that the boundary layer thickness is negligibly small compared with the dimensions of the profile holds strictly, then C_{F} represents the only contribution to the drag which is related to the viscosity of the flowing medium. If the actually existing boundary layer thickness is taken into account, then an additional contribution from viscous action is added to the drag: the *form drag*. The latter comes about as follows:

For slender profiles — and these are ordinarily encountered in hydrodynamics — the boundary layer thickness toward the tail end of the profile often grows to a magnitude comparable to the thickness of the profile

*In a strict sense, eq. (5.48) should contain the cosine of the angle between the tangent at the surface points of the profile and the direction of flow. Owing to the small thickness ratio of customary profiles, that angle is relatively small and, hence its cosine practically unity.

there (particularly for small Reynolds numbers). At sharp trailing edges of profiles, the effective profile thickness is thus equal to the sum of the displacement thicknesses on the two sides of the profile. The potential-theoretical pressure distribution of the profile, which strictly speaking predicts stagnation pressure at the trailing edge, is changed by the displacement action of the boundary layer such that a somewhat lower pressure results. Thus, the resulting pressure force in the direction of flow does not yield the value zero as predicted by potential theory. Rather there is a retarding force, the form drag, which increases strongly when the boundary layer separates from the trailing edge of the profile. Except for this extreme case of flow separation, the form drag is uniquely coupled with the boundary layer development and can be computed from it.

Pretsch [100] and Squire and Young [126] developed, almost simultaneously and independent of one another, approximation theories that allow for computing the form drag from the values of the boundary layer parameters and the velocity at the trailing edge of the profile. The central problem in both theories consists in following the boundary layer profile, which arrives at the trailing edge, through the wake until the pressure decreases to the undisturbed pressure p_∞ of the potential flow, in order to find the total momentum loss and thereby the total drag (skin friction and form drag). Both methods lead practically to the same results. While Pretsch solves this problem purely theoretically, Squire and Young employ certain empirical relations. The sum of the skin friction C_F and the form drag C_{Dr}, that is, the *profile drag**

$$c_{WP} = C_F + C_{Dr}, \tag{5.49}$$

can be represented according to both theories in the form

$$c_{WP} = \left(\frac{\delta_2}{L}\right)_S \zeta_S + \left(\frac{\delta_2}{L}\right)_D \zeta_D \tag{5.50}$$

(S = suction side, D = pressure side of the profile) where $\zeta(u_{\delta_H}/u_\infty, H)$ was given by Pretsch (Fig. 5.7), or by Squire and Young in analytic form

$$\zeta = \left(\frac{u_{\delta_H}}{u_\infty}\right)^{\frac{5+H_{12}(H)}{2}}. \tag{5.51}$$

Here u_{δ_H} is the velocity (which is generally different on the suction and pressure sides) at the trailing edge of the profile, and H_{12} is the value of the shape parameter at the trailing edge (usually different on the pressure

*In studying the lift of an airfoil of finite span it is found that the profile drag c_{WP} has to be augmented by the "induced" drag, which depends only on the projected area of the wing but not its profile and is moreover proportional to the square of the lift.

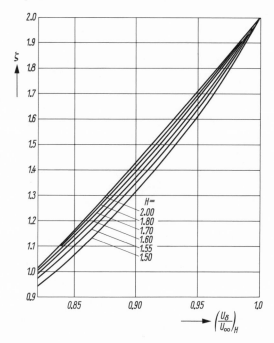

Fig. 5.7.   Computation of the profile drag $c_{\mathrm{WP}} = C_{\mathrm{F}} + C_{\mathrm{Dr}}$ according to Pretsch [100]. Function $\zeta[(u_\delta/u_\infty)_{\mathrm{H}}; H]$.

and suction sides). The velocity $u_{\delta_{\mathrm{H}}}$, however, can be determined only from potential-theoretical computation which accounts for the displacement action of the boundary layer. Practical computations have shown that an estimate for the velocity $u_\delta(x)$ in the vicinity of the trailing edge is quite sufficient for the present purpose. If, for example, $u_{\delta_{\mathrm{H}}}/u_\infty$ and thus also ζ from Fig. 5.7 is estimated too small, this will result in a greater pressure increase in the vicinity of the trailing edge and thus larger values of δ_2. The product $(\delta_2/L)\zeta$, which appears in eq. (5.50), changes very little, however. Based on measurements, it appears justified to put the value of $u_{\delta_{\mathrm{H}}}$ equal to u_δ at the point $x/L = 0.90$ as determined by potential theory. It is important to carry out the boundary layer computation also in the interval $0.9 < x/L < 1$ for the velocity distribution $u_\delta(x)$ so defined. Results for this computational method can be found in Chapter 6.

5.4.4 Computation of the maximal lift of airfoils

The maximal lift of an airfoil is reached when, under a steady increase of the angle of attack of the profile against the direction of flow, separation occurs in the domain of increasing pressure on the suction side of the profile and when the separation point moves rapidly upstream from the trailing edge as the angle of attack is further increased. The wake which develops behind the separation point influences the potential flow such that the circulation, and thus the lift, decreases strongly with increasing angle of attack.

In order to determine the maximal lift, one proceeds as follows: for a sequence of angles of attack α_g [with corresponding coefficients of lift c_a, according to eq. (5.47)] one determines the potential-theoretical velocity distribution $u_\delta(x)$, as shown in Section 5.4.2 ($x = $ coordinate along the profile surface, stagnation point at $x = 0$). The boundary layer computation of Chapter 4, with use of Sect. 5.1, yields for each distribution $u_\delta(x)$ the development of the shape parameter $H(x)$ of the boundary layer (which is usually turbulent behind the velocity maximum). It turns out that in the vicinity of the trailing edge, for example at the point $x/L = 0.9$ (which proved to be useful as a check point), the values of H decrease with increasing c_a. The value of c_a at which H is just equal to the value $H_{sep} = 1.57$ [for the lower empirical bound $H_{sep} = 1.515$, see eq. (4.41)] is equal to the maximum of c_a. Usually, $c_{a\,max}$ can be found by interpolation (see the examples in Chapter 6).

For thin profiles with a small radius at the nose, $c_{a\,max}$ may be determined by turbulent separation in the front part of the profile. In this case, the shape parameter H has a minimum in the front part of the profile which decreases with increasing values of c_a and finally drops below the separation value $H_{sep} \approx 1.57$ when $c_{a\,max}$ is reached. For example, see Fig. 6.35b, which shows a velocity distribution $u_\delta(x)$ and a shape parameter $H(x)$ corresponding to this case.

5.5 Computation of the Influence of Boundary Layer Suction by Single Slots

5.5.1 General remarks

Prandtl [94] already indicated at an early state of the development of the boundary layer theory that suction of relatively small amounts of boundary layer material into the profile would strongly influence the downstream behavior of the boundary layer. For example, it is possible to avoid flow separation and its generally highly undesirable consequences (energy losses, reduction of the lift of airplane wings) by suction of those parts of the boundary layer that are close to the wall. In laminar boundary layers it is possible to prevent transition into the turbulent state by suction, and the drag coefficients can thus be kept small. Although the technical realization of boundary layer suction poses considerable difficulties even today, it is desirable for research and planning to be able, in a given problem, to estimate what amounts of material have to be subjected to suction and what energy is necessary to achieve a certain result, as for example the prevention of flow separation for a given pressure distribution.

In the subsequent sections a simple theory will be derived which in approximation solves this problem for the most frequent case, that of suction by a single slot. This theory is based on the same simplifying concepts that were developed in the preceding chapters for the ordinary boundary layer theory. In particular, the presentation will be based on the system of equations described in Section 4.3 for Computational Method II (with *one* shape parameter for the velocity profile).

The asymptotic limiting case of continuous suction which, according to exact investigations (detailed references in [111], promises optimal utilization of a given amount of suction material, can be approximated in this theory of suction by single slots by a large number of successive slots in the direction of flow (see for example Pfenninger [88]). For simplicity's sake the theory will be presented for constant material properties only (incompressible, $\rho/\rho_\delta = 1$). The generalization to variable material properties does not present any fundamental difficulties.

The case of influencing the boundary layer by blowing out material in a direction tangential to the surface, which in many cases is just as successful as suction, particularly in preventing flow separation (see for example [21, 130]), shall not be considered here.

5.5.2 Basic approach

A typical flow pattern as observed in the vicinity of a suction slot is presented schematically in Fig. 5.8 (two-dimensional). Depending on the rate of suction, a certain fraction of the boundary layer material arriving at the slot disappears there, while the remaining material above the stagnation-point streamline (dash-dot in the figure) flows on beyond the slot. Along the stagnation-point streamline the velocity decreases from

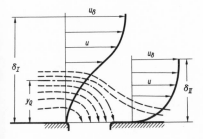

Fig. 5.8. Illustration of the flow in the vicinity of a suction slot.

the value $u(y_Q)$ at point I (leading edge of the slot) to the value zero at point II (far edge of the suction slot). The velocity profile which arrives at point I is "sliced" by the suction process at the distance y_Q from the wall. The "remaining" profile transforms itself while moving across the slot, practically without loss (an assumption made in this model), into a normal velocity profile with $y = 0$, $u = 0$ at point II. Mass, momentum, and energy are assumed to be equal for the remaining profile at point I and for the new profile arriving at point II. These assumptions lead to the following relations at point II for the layer quantities in the approximation theory presented in Section 4.3 (the subscript u indicating constant material properties, $\rho/\rho_\delta = 1$, is omitted here for clarity's sake):

$$(\delta_1)_{\text{II}} = (\delta_1)_{\text{I}} - (\delta_1)_Q, \tag{5.52}$$

$$(\delta_1)_Q = \int_0^{y_Q} \left(1 - \frac{u}{u_\delta}\right) dy, \tag{5.53}$$

$$(\delta_2)_{\text{II}} = (\delta_2)_{\text{I}} - (\delta_2)_Q, \tag{5.54}$$

$$(\delta_2)_Q = \int_0^{y_Q} \frac{u}{u_\delta}\left(1 - \frac{u}{u_\delta}\right) dy, \tag{5.55}$$

$$(\delta_3)_{\text{II}} = (\delta_3)_{\text{I}} - (\delta_3)_Q, \tag{5.56}$$

$$(\delta_3)_Q = \int_0^{y_Q} \frac{u}{u_\delta}\left[1 - \left(\frac{u}{u_\delta}\right)^2\right] dy, \tag{5.57}$$

$$H_{\text{II}} = \left(\frac{\delta_3}{\delta_2}\right)_{\text{II}}, \tag{5.58}$$

$$Z_{\text{II}} = (\delta_2 R_{\delta_2}^n)_{\text{II}}. \tag{5.59}$$

Starting with these initial values, the boundary layer calculation based on Method II can be continued beyond the suction slot to a possible second slot, where the operations (5.52) through (5.59) may be repeated for some other suction conditions. The width of the slots in all these computations is assumed small compared with the length x of the flow.

5.5.3 Details of the computation

In eqs. (5.52) through (5.57) the effects of suction are expressed by the quantity y_Q. This quantity, according to the definition in Fig. 5.8, is related to the amount of suction material Q_a (per unit length of the slot) by the relation

$$Q_a = u_\delta \int_0^{y_Q} \frac{u}{u_\delta} dy = u_\delta \delta_Q, \tag{5.60}$$

where

$$\delta_Q = \int_0^{y_Q} \frac{u}{u_\delta} \, dy \tag{5.61}$$

may be called the "suction thickness." It is customary to characterize the amount of suction material by a dimensionless coefficient

$$c_Q = \frac{Q_a}{u_\infty L}. \tag{5.62}$$

Here u_∞ is the unperturbed velocity and L a reference length, as for example the profile depth of an airplane wing.

For a given one-parametric trial solution for laminar or turbulent velocity profile u/u_δ and a given boundary layer thickness $\delta(x)$ with the shape parameter $H(x)$ as the variable, y_Q/δ is a known function of the relative suction thickness δ_Q/δ_I and the shape parameter H_I at point I:

$$\frac{y_Q}{\delta_I} = f\left(\frac{\delta_Q}{\delta_I}, H_I\right). \tag{5.63}$$

This allows to evaluate the integral expressions (5.52) through (5.58) for a given amount of suction material Q_a.

The computational method shall be explained here for *turbulent boundary layers*. For simplicity's sake, the power law

$$\frac{u}{u_\delta} = \left[\frac{y}{\delta(x)}\right]^{k(x)} \tag{5.64}$$

is chosen to represent the turbulent velocity profile. All integral expressions (5.52) through (5.58) may then be evaluated in closed form. One finds for the function in eq. (5.63):

$$\frac{y_Q}{\delta} = \left[(1 + k)\frac{\delta_Q}{\delta}\right]^{1/(1+k)}, \tag{5.65}$$

where

$$k = \frac{2 - H}{3H - 4}, \tag{5.66}$$

$$H = 2\frac{1 + 2k}{1 + 3k}. \tag{5.67}$$

In order to prevent flow separation in a turbulent boundary layer in a domain of increasing pressure, it is necessary to determine first by the

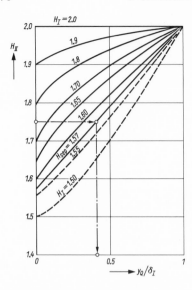

Fig. 5.9. Graph for the determination of H_{II} behind a suction slot from the value H_I before the slot for a given "suction distance" y_Q/δ_I. See also eqs. (5.36) through (5.41).

Example: $H_I = H_{sep} = 1.57$, H_{II} is expected to assume the value 1.75. The amount of suction has to be chosen such that $y_Q/\delta_I \approx 0.41$.

usual boundary layer computation from which point $x = x_{sep}$ on the danger of separation ($H \leq 1.57$) exists. The suction slot then has to be placed before x_{sep}. The problem is thus as follows: what percentage y_Q/δ of the boundary layer thickness has to be sucked away in order to achieve, for given values H_I and Z_I of the boundary layer which arrives at the slot, a "rejuvenated" boundary layer behind the slot with a given larger value H_{II}? The required "degree of rejuvenation" of the boundary layer depends upon the pressure increase which the boundary layer has to overcome behind the suction slot.

The question of the appropriate value of H_{II}, i.e., the required amount of suction, can be answered only iteratively by a sequence of boundary layer computations in which the value H_{II} is varied as a parameter. In determining the connection between y_Q/δ and $H_{II}(H_I, y_Q/\delta)$, the graphical representation in Fig. 5.9 is useful (valid for turbulent boundary layer only). It is easy to develop a similar graph for the laminar boundary layer.

An extensive description of this theory, with applications to increasing the lift of airplane wings can be found in [21], [146], [147]. An approximation theory for continuous suction is given, for example, in Eppler [23].

6 *Examples*

6.1 *Preliminary Remarks*

In order to be able to compare the accuracies of the three computational methods described, all examples, at least for the case of laminar boundary layer, will be treated by Computational Methods I, II, and III. For turbulent boundary layers (ignoring the estimating method by Buri [5], derived from method I) only Computational Method II, whose results can be compared with measurements, has any practical importance.

As the simplest example, mathematically, the boundary layer in a flow without pressure gradient, the so-called "flat-plate boundary layer," will be considered first. For the incompressible laminar and turbulent boundary layers, the solution is more or less trivial since our approximation theory in this case is based either on the exact or the empirically known solutions (cf. Chapter 3). On the other hand, the solutions for compressible flow without and with heat transfer allow for a true checking of the power of the approximation theory. The results by this approximation theory for the drag coefficient in compressible, laminar, and turbulent boundary layers are of particular importance here. This is followed — in increasing degree of difficulty — by examples of boundary layer computations in flow with pressure gradients. Special attention is given to the question of the influence of compressibility and heat transfer on the flow and temperature boundary layers, because of the increasing importance of high-speed aerodynamics. A series of examples illustrates that heat transfer in forced convection can be computed very reliably nowadays, even in the case of difficult boundary conditions (for example, when high temperature and pressure gradients are present in the direction of flow), particularly on the basis of the modified Reynolds analogy illustrated in Section 5.3.

6.2 Flows Without Pressure Gradients

6.2.1 Systems of equations

This type of flow without pressure gradient exists in the case of a flat plate in parallel flow or in that of a cylinder in axisymmetric flow, and also in the axial flow around a symmetric cone in a supersonic velocity field. In all these cases, eqs. (4.31), (4.38), and (3.62), on which the three computational methods are based, simplify because of $u'_\delta = 0$, $u_\delta = $ const., to the following:

$$Z' = F_2, \tag{6.1}$$

$$H^{*\prime} = F_4/Z, \tag{6.2}$$

$$\left[\frac{\partial}{\partial y} \left(\mu \frac{\partial u}{\partial y} \right) \right]_{y=0} = 0, \tag{6.3}$$

where

$$F_2 = (1 + n)\alpha \frac{\delta_2}{(\delta_2)_\mathrm{u}}, \tag{6.4}$$

$$F_4 = (2\beta \, \mathrm{R}_{\delta_2}^{n-N} - \alpha H^*) \frac{\delta_2}{(\delta_2)_\mathrm{u}}. \tag{6.5}$$

6.2.2 The laminar boundary layer for incompressible flow (constant material properties)

In this case, the parameters M_δ and Θ, eq. (2.160), are assumed to be zero. Thus, the compatibility condition (6.3) takes on with eq. (4.6), the form

$$Z_\mathrm{u} \frac{u'_\delta}{u_\delta} = \Gamma(H) = 0, \tag{6.6}$$

and eq. (6.2), because of (4.36), immediately becomes an equation for the determination of the shape parameter H. Moreover, with $\rho = \rho_\delta = $ const.,

$$\left(\frac{\delta_2}{(\delta_2)_\mathrm{u}} \right)_{\rho/\rho_\delta = 1} = 1, \tag{6.7}$$

and, because of $n = N = 1$ $(n - N = 0)$, one has for laminar boundary layer

$$F_2(H) = 2\alpha_l(H), \tag{6.8}$$

$$F_4(H) = 2\beta_l - \alpha_l H. \tag{6.9}$$

Computational Method I, based on eqs. (6.1) and (6.3) together with eq. (6.6), yields immediately the statement that for $u'_\delta = 0$ also $\Gamma(H) = 0$ and, by eq. (4.27), the shape parameter of the laminar boundary layer for the flat plate $H = 1.572 = \text{const.}$ (at every point x). This result corresponds to the exact Blasius solution, which is obviously a special case of the Hartree class of velocity profiles on which this theory is based.

For $H = 1.572$, eqs. (6.3) and (4.10) yield $F_2 = 2a_l = 0.441$. Thus, the solution of eq. (6.1) becomes:

$$Z_u(x) = F_2 x = 0.441x. \tag{6.10}$$

From the definition (4.4) for Z_u there follows for the momentum-loss thickness $(\delta_2)_u$, with L the length of the plate,

$$\frac{(\delta_2)_u}{L} = 0.664 \frac{\sqrt{x/L}}{\sqrt{R_L}}, \tag{6.11}$$

and for the local Reynolds number $R_{(\delta_2)u}$ based on $(\delta_2)_u$ [after multiplying eq. (6.11) by $\rho_\delta u_\delta/\mu_\delta$]

$$R_{(\delta_2)u} = 0.664\sqrt{x/L}\,\sqrt{R_L}. \tag{6.12}$$

The local drag coefficient c_{fi} can be computed from eq. (6.12) as

$$c_{fi} = 2\frac{\tau_w}{\rho_\delta u_\delta^2} = 2\frac{\alpha_l(H)}{R_{(\delta_2)u}} = 2\frac{0.221}{0.664}\frac{1}{\sqrt{\frac{x}{L}}\sqrt{R_L}} = \frac{0.664}{\sqrt{\frac{x}{L}}\sqrt{R_L}}. \tag{6.13}$$

The total drag coefficient C_{Fi} follows from eq. (6.13) by integration:[*]

$$C_{Fi} = \int_0^1 c_{fi}\,d\left(\frac{x}{L}\right) = \frac{1.328}{\sqrt{R_L}} = 2\left[\frac{(\delta_2)_u}{L}\right]_{x/L=1}. \tag{6.14}$$

The numerical factors in eqs. (6.13) and (6.14) are, as expected, in complete agreement with the values obtained from the exact solution [cf. eqs. (2.41) and (2.42)].

The ratio $\delta/(\delta_2)_u$ has the value 8.50 for the Blasius profile with $H = 1.572$, the approximation being based on eq. (3.81) and eqs. (3.86) through (3.92). With that value one finds for the total boundary layer thickness δ from (6.11) the relation

$$\frac{\delta}{L} = 5.64 \frac{\sqrt{x/L}}{\sqrt{R_L}}. \tag{6.15}$$

---

[*]From eq. (3.28) follows immediately, for $(du_\delta/dx) = 0$, $C_{Fi} = 2(\delta_2/L)_{x/L=1}$.

The usefulness of the earlier estimate for δ/L [Section 1.6, eq. (1.40)] is thus verified. The smaller numerical factor of 3.5 there, as against 5.64 here, is explained by the coarse approximation of the velocity profile by a straight line (cf. footnote on page 19).

Should a laminar boundary layer arrive at a domain without pressure gradient with an initial value of H different from 1.572, and with a certain thickness (with a value of Z_u other than 0), i.e., if the boundary layer has a "history" from a preceding domain with pressure gradient, then, in spite of these conditions, eq. (6.6) forces the value $H = 1.572$ to be achieved immediately upon entering in the domain where $u_\delta' = 0$. This is without a doubt an untenable result physically. It already indicates one limitation of Computational Method I.

Computational Method II, starting out from eqs. (6.1) and (6.2), yields in a domain without pressure gradient a constant value for H only if $H' = F_4(H)/Z_u = 0$. Analysis shows that $F_4(H)$, eq. (6.9) (with values of α_l and β_l determined from the Blasius profile), equals zero for $H = 1.572$. Computational Method II, as far as the shape parameter of the "pure" flat plate boundary layer is concerned, leads to the same (exact) result as Method I. The solution of the momentum equation (6.1) for $H = 1.572$ and the subsequent computations of the boundary layer thickness and the drag coefficients are then necessarily identical for the two methods. The superiority of Method II comes to the fore when a boundary layer with a "history" enters a domain without pressure gradient. Assume then at the point $x = x_1 = 0$, that $H_1 \neq 1.572$ and $Z_1 > 0$. For $H_1 > 1.572$ the function F_4, and thus H', is negative. $H(x)$ therefore, starting out with too large a value H_1, approaches asymptotically the value $H = 1.572$ of the pure flat plate boundary layer the faster the smaller Z_1 is. For $H_1 < 1.572$ the asymptotic approach to $H = 1.572$ takes place from below.

For $Z_1 = 0$ one finds $|F_4/Z| = |H'| = \pm\infty$, unless F_4 is put equal to zero at the same time; i.e., the correct solution $H_1 = 1.572$ is substituted into the equations. In other words: the correct similar solution $H = 1.572$ for $Z_1 = 0$ is already achieved at $x = 0$, regardless of how large (how incorrect) H_1 is chosen.

Figure 6.1 shows the result of a computation with the step formulas (4.42) and (4.43) where at the point $x_1 = 0.01$ the "incorrect" values $H_1 = 1.640$ and $Z_1 = 0.00678$ (instead of 1.572 and 0.00441) were assumed. The graph shows that the solution $H(x)$ rapidly approaches the asymptotic limiting value $H = 1.572$. If one chooses only H_1 incorrectly ($= 1.640$) but Z correctly ($= 0.00441$), one obtains the dashed curve for

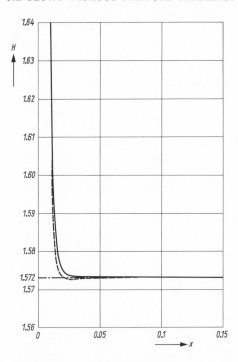

Fig. 6.1. Laminar incompressible boundary layer along a flat plate with a "history." Instead of using the correct values $H = 1.572$ and $Z_1 = 0.441 \cdot 0.01 = 0.00441$ at the point $x_1 = 0.01$, the "wrong" values, caused, e.g., by a history of $H_1 = 1.640$ and $Z_1 = 0.00648$, are assumed.

Result: $H(x)$ for

$$\left.\begin{array}{l} \text{——} \quad Z_1 = 0.00648; \\ \phantom{\text{——}} \quad H_1 = 1.640 \\ \text{---} \quad Z_1 = 0.00441; \\ \text{-·-} \quad H_1 = 1.640 \end{array}\right\} x_1 = 0.01$$

$Z_1(x) = 0.441x;$
$H = 1.572 = const.,$
exact result.

$H(x)$, which enters the domain $H < 1.572$ prior to reaching the asymptotic value.

6.2.3 The turbulent boundary layer for incompressible flow

Equations (6.1) through (6.5) may also be applied to turbulent boundary layers if only rough estimates are required. One merely has to introduce the universal functions developed for the turbulent boundary layer and put $n = 0.268$, $N = 0.168$, $n - N = 0.1$.

With *Computational Method I* it follows from eq. (6.3) that $\Gamma = 0$, and from the approximate relation for the shape parameter of the turbulent boundary layer for the flat plate, eq. (4.28), that

$$H = 1.75 = \text{const.} \tag{6.16}$$

Equation (4.14) yields with $a_t = 0.0187$ for $u_\delta = \text{const.}$

$$Z(x) = (\delta_2)_u \, \mathrm{R}_{(\delta_2)_u}^{0.268} = a_t x = 0.0187x. \tag{6.17}$$

The momentum-loss thickness $(\delta_2)_u$ and the Reynolds number $\mathrm{R}_{(\delta_2)_u}$ are then obtained, in completely analogous fashion to eqs. (6.11) and (6.12), with $n = 0.268$

$$\frac{(\delta_2)_u}{L} = a_t^{1/(1+n)} \frac{(x/L)^{1/(1+n)}}{R_L^{n/(1+n)}} = 0.0434 \frac{(x/L)^{0.788}}{R_L^{0.211}}, \tag{6.18}$$

$$R_{(\delta_2)_u} = 0.0434 \left(\frac{x}{L}\right)^{0.788} R_L^{0.788}. \tag{6.19}$$

The local drag coefficient c_{fi} can be obtained from the empirical relation (3.132), using (3.135) and (3.129). For $H = 1.75$ $[k(H) = 0.2, (\delta_1/\delta_2)_u = 1 + 2k = 1.4)]$ one has $\alpha_t = 0.0148$, and with eq. (6.19)

$$c_{fi} = 0.0868 \frac{1}{\left(\dfrac{x}{L}\right)^{0.211}} \frac{1}{R_L^{0.211}}. \tag{6.20}$$

For the total drag coefficient C_{Fi}, we find then

$$C_{Fi} = \int_0^1 c_{fi}\, d(x/L) = \frac{0.0868}{R_L^{0.211}} = 2\left[\frac{(\delta_2)_u}{L}\right]_{x/L=1}. \tag{6.21}$$

The total boundary layer thickness δ for $H = 1.75$ $(k = 0.2)$ [based on the power law (3.123)] is determined as:

$$\frac{\delta}{L} = \frac{\delta}{(\delta_2)_u} \cdot \frac{(\delta_2)_u}{L} = \frac{(1+k)(1+2k)}{k} \cdot 0.0434 \frac{(x/L)^{0.788}}{R_L^{0.211}}$$

$$= 0.365 \frac{(x/L)^{0.788}}{R_L^{0.211}}. \tag{6.22}$$

Computational Method I disregards the energy law (6.2) and the empirical dissipation law (3.142) that appears in the function F_4, eq. (6.5). The relations (6.16) through (6.22) so derived are therefore more or less coarse approximations, as will be shown.

When applying *Computational Method II* to the turbulent boundary layer, a complication arises in comparison to the laminar boundary layer, owing to the fact that the function F_4, eq. (6.5), depends also on the Reynolds number $R_{(\delta_2)_u}$ because of $n - N = 0.1$. For $M_\delta = 0$, $\Theta = 0$ we have

$$F_4 = 2\beta_t R_{(\delta_2)_u}^{0.1} - \alpha_t H. \tag{6.23}$$

The structure of F_4 thus reflects an important influence of the empirical dissipation law (3.142): we conclude from eq. (6.23) because of the

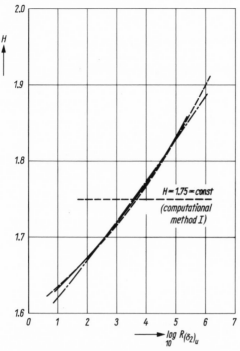

Fig. 6.2. Turbulent boundary layer along a flat plate. Functional relation $H(\mathrm{R}_{(\delta_2)_u})$.

——— Interpolation of Rotta's measurements
– – – Theory, Method II
– · – Approximation, eq. (6.25), $H = 1.572\,\mathrm{R}_{(\delta_2)_u}^{0.0132}$

dependence of $\mathrm{R}_{(\delta_2)_u}$ (and Z) on x, that no constant value of the shape parameter H exists that lets F_4 vanish at every point x. Thus, for the turbulent boundary layer along a flat (smooth) plate, there exist no similar solutions with $H = $ const. We shall see later that such similar solutions for turbulent boundary layers are possible only in flows with pressure gradients (and, according to Rotta [108], also along a flat plate, provided there is a certain x-proportional distribution of the surface roughness).

The solution of the simultaneous system of eqs. (6.1), (6.2) together with (6.23), (3.135), (3.143) results in a relation $H(\mathrm{R}_{(\delta_2)_u})$ which is illustrated in Fig. 6.2. For small Reynolds numbers $\mathrm{R}_{(\delta_2)_u} < (\mathrm{R}_{(\delta_2)_u})_{\mathrm{crit}}$ (cf. Section 5.1) we assume that the turbulent state of the boundary layer is caused by surface roughness, for example. The computational result, in the interesting $\mathrm{R}_{(\delta_2)_u}$ domain, agrees well with the empirical law

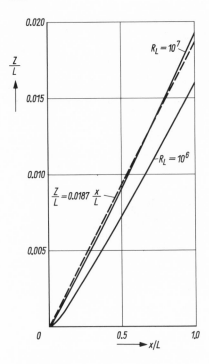

Fig. 6.3. Turbulent boundary layer along a flat plate; constant material properties. Graph of the parameter

$$Z(x) = \delta_2 \cdot R_{\delta_2}^{0.268}$$

for the boundary layer thickness with the Reynolds number R_L as a parameter.

$$R_{\delta_2} = \frac{\rho_\delta u_\delta \, \delta_2}{\mu_w} \; ;$$

$$R_L = \frac{\rho_\infty u_\infty L}{\mu_\infty}$$

——— Result with Method II
– – – Estimate with Method I
(independent of R_L)

$H(R_{(\delta_2)_u})$ given by J. Rotta (solid curve in Fig. 6.2). The law follows from the empirical relation

$$\frac{H_{12} - 1}{H_{12}} \cdot \frac{1}{\sqrt{c_{fi}/2}} = 6.3 = \text{const.,} \tag{6.24}$$

where c_{fi} derives from eq. (3.132) and $H_{12}(H)$ from eq. (3.125). In the domain $10^2 < R_{(\delta_2)_u} < 10^6$, a good approximation (cf. Fig. 6.2) is

$$H(R_{(\delta_2)_u}) = k_1 [R_{(\delta_2)_u}]^{k_2};$$
$$k_1 = 1.572, \quad k_2 = 0.0132 \quad \text{for } M_\delta = 0, \quad \Theta = 0. \tag{6.25}$$

Figure 6.2 also shows that Computational Method I with $H = 1.75 = $ const. results in an average value of H which is acceptable only for $R_{(\delta_2)_u}$ values between 10^3 and 10^4.

Figure 6.3 shows the computed values of Z/L as a function of the normalized distance x/L. Because of the dependence of the shape parameter H, eq. (6.25), on the Reynolds number, the parameter R_L appears here. The result for Computational Method I, entered here for comparison (dashed line), is independent of R_L but yields useful estimates, at least for some purposes, for Reynolds numbers around $R_L = 10^7$ (which corresponds to $10^3 < R_{(\delta_2)u} < 10^4$).

For given $Z(x)$ and $H(x)$, the local drag coefficient $c_{fi}(x)$ may then be computed. It follows from (3.132) and (6.25) that

$$c_{fi}\left(\frac{x}{L}\right) = \frac{2\alpha_t(H)}{[R_{(\delta_2)u}]^n} = 2 \cdot \frac{0.0566 H(R_{(\delta_2)u}) - 0.0842}{[R_{(\delta_2)u}]^n} \approx \frac{2\alpha}{[R_{(\delta_2)u}]^{n-\mathfrak{b}}} \cdot$$

(6.26)

For $\alpha_t(H) = \alpha_t(R_{(\delta_2)u})$ the approximation

$$\alpha_t(R_{(\delta_2)u}) \approx \alpha R_{(\delta_2)u}^{\mathfrak{b}};$$

$$\alpha = 0.00712, \quad \mathfrak{b} = 0.0870 \text{ for } M_\delta = 0, \quad \Theta = 0 \qquad (6.27)$$

was used here, which results in a maximal error of about 2 percent in the domain $3 \times 10^2 < R_{(\delta_2)u} < 3 \times 10^6$.

After substituting eq. (6.27) into (6.4) it is possible to obtain $R_{(\delta_2)u}(x/L)$ in closed form by integration from eq. (6.1). First, an elementary transformation of (6.1) yields, while observing $\rho_\delta = \rho_\infty$, $u_\delta = u_\infty$, $\mu_w = \mu_\infty$,

$$\frac{\rho_\delta u_\delta}{\mu_w} \frac{dZ}{d(x/L)} = \frac{d[R_{(\delta_2)u}]^{1+n}}{d(x/L)} = (1+n) R_{(\delta_2)u}^n \frac{dR_{(\delta_2)u}}{d(x/L)} = F_2 R_L$$

$$= (1+n)\alpha_t R_L = (1+n) \alpha R_{(\delta_2)u}^{\mathfrak{b}} \cdot R_L, \qquad (6.28)$$

$$\frac{dR_{(\delta_2 u)}}{dx/L} = \frac{\alpha_t}{[R_{(\delta_2)u}]^n} R_L = \frac{\alpha}{[R_{(\delta_2)u}]^{n-\mathfrak{b}}} R_L = \frac{c_{fi} R_L}{2}, \qquad (6.29)$$

and from there by integration

$$R_{(\delta_2)u}\left(\frac{x}{L}\right) = \left[(1+n-\mathfrak{b}) \alpha R_L \frac{x}{L}\right]^{\frac{1}{1+n-\mathfrak{b}}}. \qquad (6.30)$$

The total drag coefficient follows then by observing (6.26), (6.29) and (6.30) from the definition

$$C_{Fi}(R_L) = \int_0^1 c_{fi} \, d\left(\frac{x}{L}\right) = 2\frac{R_{(\delta_2)u}}{R_L} = \frac{\Phi}{R_L^\Psi}, \qquad (6.31)$$

where

$$\Phi = 2\left[\frac{\alpha}{1-\Psi}\right]^{1-\Psi}, \qquad (6.32)$$

$$\Psi = \frac{n-\mathfrak{b}}{1+n-\mathfrak{b}} \cdot \qquad (6.33)$$

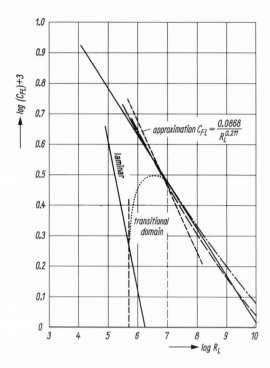

Fig. 6.4.   Total coefficient of drag $C_{\mathrm{Fi}}(R_L)$ of the turbulent boundary layer along a flat plate; constant material properties (incompressible). Comparison of different approximations with interpolation formulas for known experimental results.

—— $C_{\mathrm{Fi}} = 0.0354/R_L^{0.1533}$

– – – $C_{\mathrm{Fi}} = 0.427(\log R_L - 0.407)^{-2.64}$ (Schultz-Grunow)

– · – $C_{\mathrm{Fi}} = 0.455(\log R_L)^{-2.58}$ (Schlichting)

In the present case of incompressible flow one finds $\Phi = 0.0354$, $\Psi = 0.1533$. The law expressed by eqs. (6.31), (6.32) and (6.33), which is obtained in a semi-empirical way from the momentum and energy laws, is compared in Fig. 6.4 with known empirical laws associated with Prandtl and Schlichting [97] and with Schultz-Grunow [121] (as typical examples for many essentially equivalent laws). The agreement can be termed rather good in the interesting domain $10^6 < R_L < 10^9$. This is an indirect confirmation of the reliability of the empirical laws (3.132) and (3.142) for skin friction and dissipation. The approximation (6.21) obtained from Computational Method I and entered in Fig. 6.4 yields useful results only for $R_L \approx 10^7$, as can be expected from the result for $Z(x/L)/L$ of Fig. 6.3.

Figs. 6.5 through 6.7. Laminar compressible boundary layer with heat transfer along a flat plate.

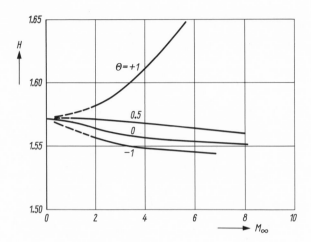

Fig. 6.5. Similar solutions $H_0(M_\infty, \Theta)$ of the system (6.1), (6.2).

$\Theta > 0$: Cooling
$\Theta < 0$: Heating
$\Theta = 1$ means $T_w = T_\delta = T_\infty$.
Assumptions: Pr $= 1$; $\mu \sim T^\omega, \omega = 1$.

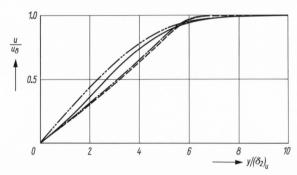

Fig. 6.6a. Velocity profile for $M_\infty = 5$, with $\Theta = 0$ (insulated wall), Pr $= 1$, $\mu \sim T^\omega, \omega = 1$.

——— Method II (Walz)
– – – Method III (Geropp)
– · – Crocco (exact)
– ·· – Blasius ($M_\infty = 0$; $\Theta = 0$)

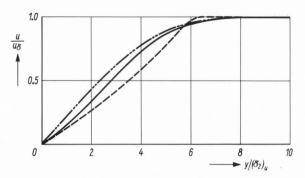

Fig. 6.6*b*.   Velocity profile for $M_\infty = 5$, with $\Theta = -1$ (Heating)
—— Method II (Walz)
– – – Method III (Geropp)
–··– Blasius ($M_\infty = 0$; $\Theta = 0$)

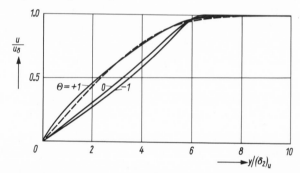

Fig. 6.6*c*.   Velocity profile for $M_\infty = 5$, with $\Theta = -1, 0, +1$, based on Method III.
—— Method III (Geropp)
– – – Blasius ($M_\infty = 0$; $\Theta = 0$)

6.2.4 The laminar boundary layer, for compressible flow with heat transfer

In this case only Computational Methods II and III will be applied and only for such boundary conditions for which exact solutions could be determined. It appears that among the known exact solutions (Busemann [6], Crocco [10], Hantzsche and Wendt [42]) the one by van Driest [17] for air as the flowing medium is the most general, since he uses the actual Prandtl number $Pr = 0.72$ and the Sutherland law in its accurate form (1.7).

Computational Method II is again based on eqs. (6.1) and (6.2). The functions F_2 and F_4, however, now depend on three parameters H, M_δ, and Θ, where $M_\delta = M_\infty$ and Θ are given with the problem (as constants). It turns out that under these assumptions, the system (6.1), (6.2) has "similar solutions" with $H = H_0(M_\infty, \Theta) = $ const., which can be determined from the zeros of $F_4(H, M_\infty, \Theta)$, Fig. 6.5. The computations for the curves displayed in Fig. 6.5 were based on the exponential law (1.8), with $\omega = 0.7$, for the dependence between viscosity μ and temperature T, which corresponds approximately to Sutherland's law as given in eq. (1.7). For other values of ω the results are similar to those of Fig. 6.5.

In Figures 6.6a, b, c, the velocity profiles are plotted which obtain for $M_\infty = 5$ for different values of the heat transfer parameter $(-1 < \Theta < 1)$ on a flat plate, based on Computational Methods II and III and the exact computations by Crocco [10] (the latter only for $\Theta = 0$). Since the latter computations are based on the viscosity law (1.8) with $\omega = 1$, Methods II and III were carried out also for $\omega = 1$ in order to achieve a valid comparison. It is for this reason that the values of $H_0(M_\infty = 5, \Theta)$ for the velocity profiles of the Hartree class (Computational Method II) can correspond only qualitatively to those in Fig. 6.5. According to all theories the velocity profiles in Figs. 6.6 become fuller with increasing Θ until finally for $\Theta = 1$, i.e., for strong cooling of the wall, the inflection point that was present at $\Theta = -1$ (heating), and still at $\Theta = 0$ (insulated wall), disappears completely (Fig. 6.6c). A velocity profile from the Hartree class that would correspond to this case with $\Theta = 1$ (Computational Method II) cannot be given because such a full profile cannot exist in the Hartree class. For comparison, Figs. 6.6a, b, c also show a plot of the well known velocity profile obtained by Blasius [4] which is based on exact calculations for $M_\infty = 0$, $\Theta = 0$.

In comparing the profiles in Fig. 6.6a we should note that the approximation solution obtained from Computational Method II selects from the Hartree class that velocity profile which by satisfying the integral conditions for momentum and energy comes closest to the exact solution. This forces a good approximation in the vicinity of the wall, so that first of all a useful approximation for the shear stress at the wall is obtained. The two-parametric approximation used in Computational Method III (Geropp) necessarily approximates the exact solution better. The temperature distribution for all velocity profiles follows from eq. (2.156) (cf. also Figs. 2.5 and 2.6).

If one substitutes the solution $H_0(M_\infty, \Theta)$ into eq. (6.1) then it follows after elementary calculations, analogous to the computations in 6.2.2, that

$$Z(x) = F_2(M_\infty, \Theta)x, \tag{6.34}$$

$$R_{\delta_2}(x) = \sqrt{R_L} \cdot \sqrt{F_2(M_\infty, \Theta)} \cdot \sqrt{x/L}, \tag{6.35}$$

$$c_f(x)\sqrt{R_L} = \sqrt{\frac{F_2(M_\infty, \Theta)}{\mu_\infty/\mu_w}} \cdot \frac{1}{\sqrt{x/L}}, \qquad (6.36)$$

$$C_F\sqrt{R_L} = 2\sqrt{\frac{F_2(M_\infty, \Theta)}{\mu_\infty/\mu_w}} = 2\sqrt{R_L}\,(c_f)_{x/L=1}. \qquad (6.37)$$

It is customary to consider the total drag coefficient C_F in relation to the value C_{Fi} (Section 6.2.2) obtained from incompressible flow (Fig. 6.7). One observes that C_F/C_{Fi} decreases with increasing Mach number. Cooling of the wall ($\Theta > 0$) diminishes the rate of decrease of M_∞, heating (transfer of heat into the boundary layer) increases the same. Knowing the dependence of the viscosity of gases on the temperature [μ increases with T, cf. eq. (1.7)], we would expect the opposite behavior of C_F/C_{Fi} as a function of M_∞ and Θ. The change of the density ρ in the function $\delta_2/(\delta_2)_u$ of eq. (6.4) (for example, ρ decreases rapidly with increasing Mach number M_∞), and the change in the velocity profile (for example, the increase of H_0 and $\alpha_l(H_0)$ and thus of C_F in the case of cooling, $\Theta > 0$, cf. Fig. 6.5), however dominates over the opposite influence by μ.

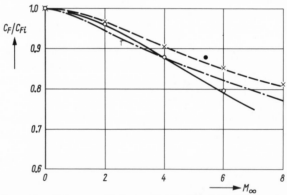

Fig. 6.7.   The ratio of drag coefficients $C_F/C_{Fi}(M_\infty, \Theta)$. Comparison with the exact solutions by van Driest [17] and Hantzsche and Wendt [42] as well as Geropp's solution [33] (Method III).

—— exact (van Driest) ⎫
–·– Method II (Walz) ⎬ Sutherland law for $\mu(T)$
O   Method III (Geropp)⎭

--- exact (Hantzsche and Wendt), Pr $= 1$; $\omega = 0.8$ ⎫
×   Method III (Geropp),   Pr $= 1$; $\omega = 0.8$ ⎬ $\mu \sim T^\omega$
●   exact (Hantzsche and Wendt), Pr $= 0.7$; $\omega = 0.8$ ⎭
        and Method III (Geropp),   Pr $= 0.7$; $\omega = 0.8$

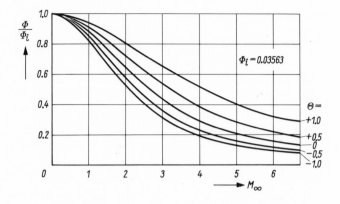

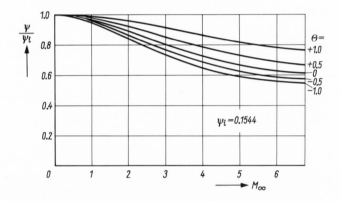

Figs. 6.8 and 6.9.   Semiempirical functions $\Phi(M_\infty, \Theta)/\Phi_i$ and $\Psi(M_\infty, \Theta)/\Psi_i$ according to eqs. (6.32), (6.33), and (6.38) used in the analytic representation of the drag coefficient $C_F(R_L, M_\infty, \Theta)$ for turbulent boundary layers with heat transfer along a flat plate.

Figure 6.7 contains plots for $\Theta = 0$ (heat-insulated wall) of the exact solution by van Driest [17] for Pr = 0.72 and μ according to eq. (1.7), as well as the exact solutions by Hantzsche and Wendt [42], together with the results obtained from Computational Method III with different Prandtl numbers and viscosity laws.

Computational Method III achieves practically (within graphical accuracy) full agreement with the exact solution. Computational Method II gives satisfactory results up to $M_\infty \approx 5$ (the maximal error in this domain is less than 2 percent). For $\Theta = 1$ (cooling, so that $T_W = T_\delta$), the error is even somewhat smaller. For heating, $\Theta < 0$, which can hardly be expected in the practice of supersonic flow, no exact results are available for comparison.

The values $+1$ and -1 for the parameter Θ are arbitrarily chosen limits in the numerical computations, limits which are expected, however, to cover the interesting range of values.

6.2.5 The turbulent boundary layer, for compressible flow with heat transfer

The method of solution indicated in Section 6.2.3 yields in this case the shape parameter H_0 as a function of R_{δ_2}, M_∞, and Θ. In the approximations for H_0 and $\alpha_t \delta_2/(\delta_2)_u$, which correspond to eqs. (6.25) and (6.27), k_1, k_2, $\mathfrak{a}$, $\mathfrak{b}$, are functions of the parameters M_δ and Θ. After integration one finds for C_F the analytical representation

$$C_F(R_L, M_\infty, \Theta) = \frac{\Phi(M_\infty, \Theta)}{R_L^{\Psi(M_\infty, \Theta)}} \left(\frac{\mu_w}{\mu_\infty}\right)^{\Psi(M_\infty, \Theta)} \tag{6.38}$$

with Φ and Ψ defined by eqs. (6.32) and (6.33). Figures 6.8 and 6.9 show the graphs of

$$\Phi(M_\infty, \Theta)/\Phi_i; \quad \Phi_i = \Phi(M_\infty = 0, \Theta = 0), \tag{6.39}$$

and

$$\Psi(M_\infty, \Theta)/\Psi_i; \quad \Psi_i = \Psi(M_\infty = 0, \Theta = 0). \tag{6.40}$$

Figure 6.10 compares $C_F(M_\infty, \Theta, R_L)/C_{Fi}$ with available measurements. It is worth noting that the theory predicts an increase of C_F/C_{Fi} (although small, because of the small power $\Psi - \Psi_i$) with the Reynolds number R_L, while the measurements show no unequivocal influence of R_L.

The influence of heat transfer at constant Reynolds number $R_L = 10^7$ can be seen in Fig. 6.10. As in the case of laminar boundary layer, C_F/C_{Fi} increases when the wall is cooled ($\Theta > 0$) and decreases in the case of a heated wall ($\Theta < 0$). This theoretical trend of the influence of Θ agrees with the available measurements only at Mach numbers up to about 5 [see the measurements by Sommer-Short for $T_W = T_\delta$ ($\Theta = 1$), $R_L = 3 \cdot 10^6$]. It is not clear at the present time whether it is the uncertainties in the rather difficult measurement techniques at high Mach numbers or inefficiencies of the theory that are responsible for the observed discrepancies with the local drag coefficient c_f (see the critical remarks by Morkovin and Walz in [72], pp. 367–380, and 299–352; also Rotta [109]).

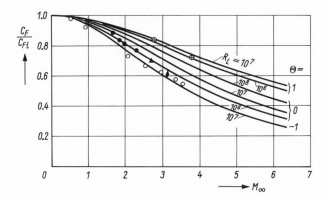

Fig. 6.10. Ratio of the total drag coefficients $C_F(R_L, M_\infty, \Theta)/C_{Fi}$. Results of Method II with the assumptions: Recovery factor $r = 0.88$; dissipation law (3.158) with (3.143) and (3.144); $H_{12}(H)$ from eq. (3.125). Local drag coefficient from eq. (3.131) with (3.135).

References to these measurements, see [72], article by Walz, pp. 299–352.

—— Approximation theory from Section 6.25.

Measurements (weighings):

O   Chapman/Kester $R_L = 6$ to 16×10^6
▲   Rubesin/Maydew/Varga $R_L = 7 \times 10^6$
●   Wilson $R_L = 10^7$ $\Bigg\}\ \Theta = 0$
♦   Brinich/Diaconis $R_L = 3$ to 18×10^6
⊕   Sommer/Short $R_L = 3 . 10^6$; $\Theta = 1$

6.2.6 The axisymmetric boundary layer around a circular cone in supersonic axial flow

In this flow geometry constant velocity exists along the generatrices of the circular cone, $u_\delta(x) = $ const.* The computation of the laminar and turbulent boundary layers is again based on the simultaneous system of eqs. (6.1), (6.2), (6.3), under the rules established in Section 5.2 for axisymmetric boundary layers.

This example is intended to show how Mangler's coordinate transformation (5.10), (5.11) points out simple connections between the drag coefficients c_f, C_F of the circular cone and the values of $\bar{c}_f$, $\bar{C}_F$ for the flat plate. The flow along the flat plate turns out to be the plane substitute

---

*We assume here that the cone angle is smaller than the maximal angle of deflection that corresponds to the Mach number of the unperturbed flow (cf., for example, J. Zierep [160], Chapter 8).

problem, in the sense of Mangler's theory, for the supersonic flow along a circular cone, where $\bar{u}_\delta = u_\delta$, $\overline{M}_\delta = M_\delta$, $\overline{\Theta} = \Theta$, $\overline{T}_\delta = T$, $\bar{\rho}_\delta = \rho_\delta$.

In the axisymmetric case the x axis is assumed to coincide with the generatrix of the circular cone. The vertex of the cone is associated with the leading edge $x = 0$ of the flat plate with length L. The cross-sectional radius $\Re(x)$ as a function of x is given by the linear relation

$$\frac{\Re}{L} = k \frac{x}{L},$$
(6.41)

where k is a function of the cone angle. The point x on the circular cone, is then associated by eq. (5.10) with the point

$$\frac{\bar{x}}{L} = \int_0^{x/L} \left(k \frac{x}{L}\right)^{1+n} \mathrm{d} \frac{x}{L} = \frac{k^{1+n}}{2+n} \left(\frac{x}{L}\right)^{2+n},$$
(6.42)

where $n = 1$ for laminar and $n = 0.268$ for turbulent boundary layer. For lengths in y direction, as for example the momentum loss thickness δ_2, we have, according to eq. (5.11), the transformation

$$\bar{\delta}_2 = \frac{\Re}{L} \delta_2 = k \frac{x}{L} \delta_2.$$
(6.43)

The shape parameter H in the laminar case is exactly $H = \overline{H} = \text{const.}$ (similar solution, as in Section 6.2.4). For turbulent boundary layers $H(R_{\delta_2}, M_\delta, \Theta)$ and $\overline{H}(\overline{R}_{\delta_2}, \overline{M}_\delta, \overline{\Theta})$ are practically equal at the points x and $\bar{x}$, which are associated with one another by eq. (6.42), because of the small variation of H with R_{δ_2}.

In the $\bar{x}$, $\bar{y}$ plane the local drag coefficient is defined as

$$(\bar{c}_f)_{\bar{x}} = 2 \frac{\alpha(\overline{H})}{\overline{R}_{\delta_2}^n} \frac{\bar{\delta}_2}{(\bar{\delta}_2)_u}.$$
(6.44)

At the point x on the generatrix of the circular cone, which corresponds by eq. (6.42) to the point $\bar{x}$ on the flat plate, we find from eq. (3.131) for the local drag coefficient of the axisymmetric system, observing eq. (6.43),

$$(c_f)_x = 2 \frac{\alpha(H) \dfrac{\delta_2}{(\delta_2)_u}}{(R_{\delta_2}^n)_x} = 2 \frac{\alpha(H) \delta_2/(\delta_2)_u}{(\overline{R}_{\delta_2}^n)_x (\delta_2/\bar{\delta}_2)^n} = 2 \frac{\alpha(H)}{(\overline{R}_{\delta_2}^n)_x} \left(\frac{kx}{L}\right)^n \cdot \frac{\delta_2}{(\delta_2)_u}.$$
(6.45)

The values $(\bar{c}_f)_x$ and $(c_f)_x$ are now to be compared at equal distances from the leading edge on the flat plate and the circular cone, that is, for the case $\bar{x} = x$. To this end it is necessary to determine first how $(\overline{R}_{\delta_2})_{\bar{x}}$ at the point $\bar{x}$ relates to $(\overline{R}_{\delta_2})_x$ at the point x. According to eqs. (6.18) and (6.19), $\overline{R}_{\delta_2}$ and R_{δ_2} vary with $\bar{x}^{1/(1+n)}$ and $x^{1/(1+n)}$ (see also the preceding footnote). Thus, we have

$$\left[\frac{(\overline{R}_{\delta_2})-}{(\overline{R}_{\delta_2})_x}\right]^n = \left(\frac{\bar{x}}{x}\right)^{n/(1+n)} = \frac{(kx/L)^n}{(2 + n)^{n/(1+n)}}, \tag{6.46}$$

and $(\bar{c}_f)_x$ becomes [at a distance x instead of $\bar{x}$, where $\delta_2/(\delta_2)_u \approx \bar{\delta}_2/(\bar{\delta}_2)_u$ because of $H \approx \overline{H}$]

$$(\bar{c}_f)_x = \frac{2\alpha(H)}{(\overline{R}_{\delta_2}^n)_x} \cdot \left(\frac{\bar{x}}{x}\right)^{n/(1+n)} \cdot \frac{\delta_2}{(\delta_2)_u}, \tag{6.47}$$

and finally

$$\frac{(c_f)_x}{(\bar{c}_f)_x} := (2 + n)^{n/(1+n)}. \tag{6.48}$$

For the *laminar* boundary layer with $n = 1$, eq. (6.48) leads to the relation

$$\frac{(c_f)_x}{(\bar{c}_f)_x} = \sqrt{3}, \tag{6.49}$$

which was already found by Hantzsche and Wendt [42].

For *turbulent* boundary layer with $n = 0.268$ one obtains

$$\frac{(c_f)_x}{(\bar{c}_f)_x} = 2.268^{\frac{0.268}{1.268}} = 1.188. \tag{6.50}$$

Integration over the surface of the cone leads to the following expression for the total drag coefficient:

$$\frac{C_F}{\overline{C}_F} = \frac{2}{(2 + n)^{1/(1+n)}} (\cos \vartheta)^{n/(1+n)}. \tag{6.51}$$

Here C_F is related to the surface of the cone and the stagnation pressure of the velocity along the generatrix, and ϑ is the half-angle of the cone. For $\vartheta = 20°$ one finds $(\cos \vartheta)^{n/(1+n)}$ for the laminar boundary layer to be 0.97, and for the turbulent boundary layer 0.99. For cone angles $\vartheta < 20°$, the factor $(\cos \vartheta)^{n/(1+n)}$ in eq. (6.51) may be put equal to unity. We then have

$$\frac{C_F}{\overline{C}_F} \approx \frac{2}{(2 + n)^{1/(1+n)}} \qquad (\vartheta < 20°). \tag{6.52}$$

From this equation follows for the *laminar* boundary layer with $n = 1$

$$\frac{C_F}{\overline{C}_F} = \frac{2}{\sqrt{3}} = 1.154, \tag{6.53}$$

and for the *turbulent* boundary layer with $n = 0.268$

$$\frac{C_F}{\overline{C}_F} = \frac{2}{2.268^{0.788}} = 1.050. \tag{6.54}$$

The smaller numerical values for the ratio of the total drag coefficients as compared to the ratio of the local drag coefficients (for example $C_F/\overline{C}_F = 1.154$, as compared to $(c_f/\overline{c}_f)_x = \sqrt{3} = 1.73$ for laminar boundary layer) can be explained by the observation that in the computation of C_F by integration over the local drag coefficient c_f the large values of c_f in the vicinity of the vertex of the cone (in the vicinity of $x = 0$) are to be multiplied by small elements of area that approach zero in the limit when $x = 0$. Therefore, C_F for the cone should have a smaller average than for the flat plate where c_f has to be multiplied with the same element of area $1 \cdot dx$ at every point x.

Since for the turbulent boundary layer the shape parameter H depends (very slightly) on R_{δ_2} according to eq. (6.25), it is, strictly speaking, necessary to replace in all equations in this section, in the case of the turbulent boundary layer, n by $n - \mathfrak{b}$, with $\mathfrak{b}(M_\delta, \Theta)$ given by Section 6.2.5. Equation (6.50) then yields for $M_\delta = 5, \Theta = 0$ the value 1.083 instead of 1.188. Inasmuch as this section is intended to give merely an orientation about the orders of magnitude of the differences between the drag coefficients for the cone and the flat plate, the given relations should be sufficiently accurate.

6.3 Special Flows with Pressure Gradients Leading to Closed-Form Solutions

6.3.1 General remarks

Equations (4.31), (4.38), and (3.62) in their general form, are chosen as a basis in this case:

$$Z' + Z\frac{u'_\delta}{u_\delta}F_1 - F_2 = 0 \qquad (\text{for } T_w = \text{const.}, \mu'_w = 0), \tag{6.55}$$

$$H^{*\prime} + H^*\frac{u'_\delta}{u_\delta}F_3 - \frac{F_4}{Z} = 0, \tag{6.56}$$

$$\left[\frac{\partial}{\partial y}\left(\mu\frac{\partial u}{\partial y}\right)\right]_{y=0} = \frac{dp}{dx} \tag{6.57}$$

where F_1 through F_4 are given by eqs. (4.32), (4.34), (4.39), and (4.40), respectively. We shall first treat examples that lead to a closed-form solution of the simultaneous systems of eqs. (6.55), (6.57) or (6.55), (6.56) without simplifying assumptions, such as piecewise linearity of $u_\delta(x)$ according to eq. (4.15). We shall see that these examples will lead to the so-called similar solutions, which are characterized by a constant shape parameter H.

6.3.2 Similar solutions of the laminar boundary layer; incompressible case

We consider first the solution of the pair of eqs. (6.55), (6.57) based on a one-parametric trial solution of the type (3.79) with the unknowns Z and H. For incompressible flow, eq. (6.57) has the form

$$Z\frac{u'_\delta}{u_\delta} = -\left[\frac{\partial^2 u/u_\delta}{\partial[y/(\delta_2)_u]^2}\right]_{y=0} = \Gamma(H). \tag{6.58}$$

Substituting (6.58) into (6.55) yields for $Z(x)$ the solution

$$Z(x) = \int_{x_1}^{x} F(H)\,dx + Z_1, \tag{6.59}$$

where

$$F(H) = \Gamma \cdot F_1 - F_2, \tag{6.60}$$

$$Z_1 = Z(x_1). \tag{6.61}$$

Similar solutions, by assumption, are given when $H = $ const. and thereby also $F(H) = $ const. It then follows from eqs. (6.60) and (6.61):

$$Z(x) = F x + Z_1, \tag{6.62}$$

which is a linear relation between Z and x. Using eq. (6.62), one obtains from (6.58) first

$$\frac{du_\delta}{u_\delta} = \frac{\Gamma}{F x + Z_1}\,dx = \frac{\Gamma}{F}\frac{d(F x + Z_1)}{F x + Z_1} \tag{6.63}$$

and from here by integration

$$u_\delta(x) = C(F x + Z_1)^m \tag{6.64}$$

($m = \Gamma/F = $ const.; $C = $ integration constant). If we put $Z = Z_1 = 0$ for $x = 0$, we find

$$u_\delta \sim x^m \tag{6.65}$$

to be the type of velocity distribution $u_\delta(x)$ which leads to solutions $H = $ const.

This result is as yet completely independent of the choice of the one-parametric trial solution for u/u_δ. It is remarkable that this very simple form of approximation theory by von Kármán [56] and Pohlhausen [91], on which Computational Method I is based, should already lead to the power law (2.27) of the exact theory of similar solutions. If one takes for u/u_δ the trial solutions (3.81) and (3.86) through (3.92) which interpolate the exact solutions of Hartree [43], then the approximation theory based on eqs. (6.55), (6.57) is a complete substitute for the exact theory so long as $u_\delta(x)$ is of the power-law type (6.65).

The separation profile with the tangent at the wall $(\partial u/\partial y)_{y=0} = 0$ ($\alpha = 0, F_2 = 0$) as a similar solution occurs for the exponent

$$m_{\text{sep}} = -\frac{1}{(F_1)_{\text{sep}}} = -\frac{1}{3 + 2 \cdot 4.038} = -0.0904 \qquad (6.66)$$

(in complete agreement with the exact solution).

The result of this approximation theory becomes, however, as we shall see in Sect. 6.4, the less certain the more $u_\delta(x)$ differs from the power-law type (6.65). Starting from a stagnation point in whose neighborhood $u_\delta \sim x$, this simple theory (Computational Method I) lets us expect always reliable results for accelerated flows.

If one starts out from the system of equations (6.55), (6.56) (the base for Computational Method II), then the question of similar solutions with $H^* = H = \text{const.}, H' = 0$ in eq. (6.56) leads first to

$$Z\frac{u_\delta'}{u_\delta} = \frac{F_4}{H F_3} = \text{const.}, \qquad (6.67)$$

and with Z, according to eq. (6.62) (for $Z_1 = 0$) assumed proportional to x, again to

$$u_\delta \sim x^m.$$

Here the exponent becomes

$$m = \frac{F_4}{H F_2 F_3 - F_1 F_4}, \qquad (6.68)$$

(which is negative for retarded flow). Again this approximation theory based on eqs. (6.55) and (6.56) turns out to be identical with the exact theory if the trial solution u/u_δ is taken to be the exact Hartree solution [eqs. (3.81) with (3.86) through (3.92)) and $u_\delta(x) \sim x^m$. The separation profile with $\alpha = 0, F_2 = 0$ occurs when

$$m_{\text{sep}} = -\frac{1}{(F_1)_{\text{sep}}} = -0.0904, \qquad (6.69)$$

in agreement with (6.66).

In computed examples we shall show later that deviations of $u_\delta(x)$ from the power-law type lead nevertheless to acceptable agreement with exact solutions, contrary to the approximation theory by von Kármán and Pohlhausen [91] (cf. Section 6.4.1, Fig. 6.12a). One can conclude that the energy law (6.56) has more predictive value physically than the compatibility condition (6.57).

6.3.3 Similar solutions of the turbulent boundary layer; incompressible case

According to Section 3.5.3.4, turbulent velocity profiles should be approximated by at least two shape parameters, namely H, according to eq. (3.75), and R_{δ_2}, according to eq. (3.74). The shape parameter H has dominant influence, even more so in incompressible flow. Similarity of the turbulent velocity profiles which develop in x direction is therefore strictly speaking only realized when, together with $H = $ const., also $R_{\delta_2} = $ const. (or the local drag coefficient $c_f = $ const.). The second condition can obviously not be satisfied, unless (following Rotta [108]) one provides for variations in the roughness along the surface. Rotta proposed, based on theoretical consideration, to characterize turbulent velocity profiles in incompressible flows by the parameter

$$J(H, c_{fi}) = \frac{H_{12} - 1}{H_{12}} \cdot \frac{1}{\sqrt{c_{fi}/2}} = \frac{H_{12} - 1}{H_{12}} \frac{R_{(\delta_2)u}^{n/2}}{\sqrt{\alpha_t}} \qquad (6.70)$$

(where c_{fi} is taken from eq. (3.132), for smooth walls) and to define similar solutions by $J = $ const. According to Clauser [8] one may also consider the quantity

$$\Pi = \frac{(\delta_1)_u}{\tau_w} \cdot \frac{dp}{dx} = -(\delta_2)_u R_{(\delta_2)u}^n \frac{du_\delta/dx}{u_\delta} \frac{H_{12}}{\alpha_t} = -Z \frac{u_\delta'}{u_\delta} \frac{H_{12}}{\alpha_t} \qquad (6.71)$$

as a characteristic boundary layer parameter and to require for similar turbulent boundary layers $\Pi = $ const. There is a one-to-one correlation between J and Π which can be approximated rather well by

$$J \approx 6.1 + 1.9\Pi \quad (-1 < \Pi < 10). \qquad (6.72)$$

(This correlation was used in determining the empirical law for dissipation, eqs. (3.147) through (3.150); cf. also Section 3.7.3.)

Because of the moderate influence of $R_{(\delta_2)u}$ on the velocity profile (the order of magnitude of this influence can be seen in Fig. 6.2 for the flat plate), it is possible to define similar turbulent boundary layers also by $H = $ const. within the framework of the approximation theory developed here. Conditions for such solutions will be considered now. The minor influence of the Reynolds number $R_{(\delta_2)u}$ on the solutions will be recognizable.

We take as a starting point the system of eqs. (6.55) and (6.56). With $H' = 0$, $H = $ const., it follows from eq. (6.56) (as in the case of laminar boundary layers)

$$Z \frac{u'_\delta}{u_\delta} = \frac{F_4(H, R_{(\delta_2)u})}{H\,F_3(H)}.$$ (6.73)

Substitution into eq. (6.55) yields

$$Z' = F_2 - \frac{F_4\,F_1}{H\,F_3} = F_t(H, R_{\delta_2}).$$ (6.74)

The small exponent which occurs with the x-dependent quantity $R_{(\delta_2)u}$ in F_4 indicates that $R_{(\delta_2)u}$ will have little influence on the solution sought after. We therefore put, on a trial basis, $R_{(\delta_2)u} = $ const. This is equivalent to saying that $R_{(\delta_2)u}$ will occur as a parameter in the solution. But for $F_t = $ const. we have, as in the laminar case, $Z = $ const. $\cdot x$. Substitution into eq. (6.73) and subsequent integration leads, as in the case of the laminar boundary layer, to the solution

$$u_\delta(x) \sim x^m,$$ (6.75)

where

$$m(H, R_{(\delta_2)u}) = \frac{F_4}{H\,F_2F_3 - F_1F_4}.$$ (6.76)

Thus, also for turbulent boundary layer, we can expect similar solutions (defined by $H = $ const., $R_{(\delta_2)u} = $ const.) for free-stream velocities $u_\delta(x)$ which vary with a power of x. This is in agreement with results of Rotta [108] and Clauser [8].

We set out to compute the (negative) value of m that corresponds to a retarded flow which, at every point x, generates the velocity profile characteristic of turbulent separation, with a wall tangent $\alpha_t(H) = 0$, $(F_2 = 0)$. From eq. (6.76) there follows for $F_2 = 0$, in formal agreement with the result for the laminar boundary layer.

$$m_{\text{sep}} = -\frac{1}{(F_1)_{\text{sep}}} = -\frac{1}{2 + n + (1 + n)H_{12_{\text{sep}}}}.$$ (6.77)

Observations show that H_{sep} may vary between 1.50 and 1.57 and thus, from the empirical law for $H_{12}(H)$, eq. (3.125), $H_{12_{\text{sep}}}$ varies between

2.70 and 2.0. Thus, m_{sep} is confined to the interval

$$-0.176 > m_{sep} > -0.208 \tag{6.78}$$

or approximately

$$m_{sep} \approx -0.19. \tag{6.79}$$

Stratford [128] and Townsend [137] found (semi-empirically)

$$m \approx -0.23. \tag{6.80}$$

This value is about twice that for laminar boundary layer (-0.0904). This comparison indicates that the turbulent boundary layer can withstand larger pressure increases than a laminar boundary layer. Later examples will confirm this fact.

In order to prevent, with certainty, separation in a practical case of retarded flow of the type $u_\delta \sim x^m$, it is necessary to make sure that $H > H_{sep}$. If one requires, for example, $H = 1.60 = \text{const.}$, one can determine the corresponding exponent m from eq. (6.76). Since the function $F_4(H, R_{(\delta_2)_u})$ remains in eq. (6.76) when $F_2 \neq 0$, we obtain m as a function of the Reynolds number $R_{(\delta_2)_u}$. From Table 6.1 we see that m becomes more negative with increasing $R_{(\delta_2)_u}$.

Table 6.1

$R_{(\delta_2)_u}$	0	10^2	10^3	10^4	10^5	10^6	∞
m	-0.066	-0.167	-0.183	-0.194	-0.200	-0.204	-1

Example of a power-law velocity distribution $u_\delta(x) \sim x^m$: $m(R_{\delta_2})$ is to be determined such that the shape parameter $H = 1.60 = \text{const.}$ For $H = 1.60$ we have $H_{12} = 1.82$ [according to eq. (3.125)]; $\alpha_t = 0.0064$; $\alpha_t H = 0.01025$; $F_1 = 2.268 + 1.628 \cdot H_{12} = 4.578$; $F_3 = 1 - H_{12} = -0.82$; $F_4 = 0.0112 R_{(\delta_2)_u}^{0.1} - \alpha_t H$. Thus, we have

$$m(R_{(\delta_2)_u}) = \frac{F_4}{-0.01065 - 4.578 F_4}. \tag{6.81}$$

In the interesting region of Reynolds numbers $10^3 < R_{(\delta_2)_u} < 10^6$, the influence of $R_{(\delta_2)_u}$ on the result of the computation is not expected to be very large. In the limiting case $R_{(\delta_2)_u} \to \infty$, and thus $m \to -1$, the empirical foundation of the theory appears overstressed.

The general numerical evaluation of $m(H, R_{(\delta_2)_u})$ according to eq.

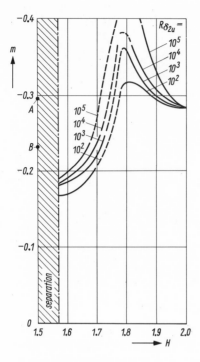

Fig. 6.11. Similar solutions for incompressible turbulent boundary layers in retarded free-stream flows of type $u_\delta \sim x^m$. The exponent m is plotted as a function of the shape parameter H with $R_{(\delta_2)_u}$ as a parameter for the family of curves. The domain with dashed curves indicates numerical uncertainties in the empirical functions F_1 through F_4.

A = maximal value of m according to Clauser
$B = m_{\mathrm{sep}}$, according to Townsend and Stratford.

(6.76) is displayed in Fig. 6.11. There is an indication of a maximum for $-m$ which depends on $R_{(\delta_2)_u}$, a fact also observed experimentally by Clauser [8]. The numerical evaluation of eq. (6.76) is, however, uncertain just in the vicinity of this maximum of $-m$, apparently because of the difference of empirical quantities in the denominator of eq. (6.76). This domain of numerical uncertainty is indicated by the dashed line in Fig. 6.11.

6.3.4 Similar solutions for compressible laminar and turbulent boundary layers

This problem is principally solvable on the basis of the system (6.55), (6.56). One again puts $H' = 0$, $H = $ const. and solves the system for $u_\delta(x)$. But since in this case the universal functions H^*, F_1 through F_4 also depend on the Mach number M_δ and thus on u_δ, it is not possible to integrate in closed form. Fernholz [31, 32] gave a numerical solution for the case $M_\delta < 1$ and $H = 1.57 = H_{\mathrm{sep}} = $ const. Solutions of the approximation theory for higher Mach numbers, with and without heat transfer, which would allow a comparison with the similar solutions for laminar boundary layer given by Li and Nagamatsu [66] and Geropp [34] have yet to be obtained.

6.3.5 Effect of a discontinuity in the velocity (pressure jump) on the boundary layer

This problem is presented, for example, in the neighborhood of a plane shock wave* and can be treated with the system (6.55), (6.56). For very short lengths of the boundary layer in strong pressure gradients one may neglect the frictional terms F_2 in eq. (6.55) and F_4/Z in eq. (6.56) against the inertial and pressure forces.† The system of equations to be solved is therefore as follows:

$$\frac{dZ}{Z} = -F_1 \frac{du_\delta}{u_\delta}, \tag{6.82}$$

$$\frac{dH^*}{H^*} = -F_3 \frac{du_\delta}{u_\delta}. \tag{6.83}$$

It is possible to solve the system (6.55) and (6.56) in closed form in the case of *incompressible* flow, both for laminar and turbulent boundary layers, based on the analytical expressions (4.32), (4.39), and (3.125) for $F_1(H)$ and $F_3(H)$. According to Fernholz [31, 32], the following results obtain for *laminar* boundary layer (subscript 1 for values prior to the jump, subscript 2 after the jump):

$$\frac{(u_\delta)_1}{(u_\delta)_2} = \left(\frac{H_1}{H_2}\right)^{0.3296} \cdot e^{0.6455(H_1-1.519)^{1.55}}, \tag{6.84}$$

$$\frac{(\delta_2)_2}{(\delta_2)_1} = a\left(\frac{H_2}{H_1}\right)^{1.6586}; \tag{6.85}$$

$$\ln a = 3.250(H_1 - 1.519)^{1.55},$$
$$+ 0.2458(H_1 - 1.519)^{1.288},$$
$$- 1.818(H_1 - 1.519)^{1.838}. \tag{6.86}$$

For *turbulent* boundary layer, [with the power law (3.123) for u/u_δ]:

$$\frac{(u_\delta)_1}{(u_\delta)_2} = \frac{H_2}{H_1}\sqrt{\frac{2 - H_2}{2 - H_1}}, \tag{6.87}$$

$$\frac{(\delta_2)_2}{(\delta_2)_1} = \left(\frac{H_2}{H_1}\right)^2\left(\frac{2 - H_2}{2 - H_1}\right)^{1.5}. \tag{6.88}$$

*The present computational method for boundary layers based on eqs. (6.55) and (6.56) can be applied for *weak* shock waves only. In *strong* shock waves, the boundary layer simplifications (2.71) and (2.72), according to Prandtl, are not valid.
†This conclusion follows from the quadrature formulas (4.42) and (4.43) if the step length Δx is chosen very small.

The largest velocity jump that can occur without leading to separation can be determined directly from eqs. (6.84) and (6.87). It is also seen that the value of H_1 just prior to the jump plays an important role. We assume for the subsequent computations that $u_\delta(x)$ was constant prior to the velocity jump. For laminar boundary layer this means $H_1 = H_1(R_{(\delta_2)_u})$, according to eq. (6.25). We now determine the ratio of the velocities $(u_\delta)_1/(u_\delta)_2$ at the jump for the case when $H_2 = H_{sep}$. We thus put in the laminar case $H_2 = 1.519$,* and in the turbulent case $H_2 = 1.57$.

The computation with eq. (6.84) yields for *laminar* boundary layer

$$\frac{(u_\delta)_1}{(u_\delta)_2} = 1.010. \tag{6.89}$$

For *turbulent* boundary layer, the result contains the Reynolds number $R_{(\delta_2)_u}$ as a parameter since H_1 is a function of that Reynolds number (see Table 6.2). One recognizes that in the case of laminar boundary layer, for the chosen initial condition $H_1 = 1.572$, a jump in the velocity of only about 1 percent is sufficient to achieve laminar separation. For turbulent boundary layer, on the other hand, considerably larger velocity jumps (pressure jumps) are permissible. H_1 increases with increasing Reynolds number $R_{(\delta_2)_u}$, as does the permissible velocity jump.

Table 6.2

$(R_{(\delta_2)_u})_1$	10^2	10^3	10^4	10^5	10^6
H_1	1.670	1.717	1.768	1.830	1.902
$(u_\delta)_1/(u_\delta)_2$	1.075	1.127	1.207	1.364	1.730

6.4 Retarded Flow Near a Stagnation Point, $u_\delta/u_\infty = 1 - x/L$

6.4.1 The Laminar incompressible boundary layer

For this example, the exact solution by Howarth [49] is available for comparison with the approximate solutions. A closed form for the approximate solution in this case is only possible with Computational Method I. Assuming $x_1 = 0$, $Z_{u1} = 0$, it follows from eq. (4.13) that

$$Z_u(x) = a_l \frac{\int_0^{x/L} \left(1 - \frac{x}{L}\right)^{b_l} d\frac{x}{L}}{\left(1 - \frac{x}{L}\right)^{b_l}} = \frac{a_l}{b_l + 1} \cdot \frac{1 - \left(1 - \frac{x}{L}\right)^{b_l+1}}{\left(1 - \frac{x}{L}\right)^{b_l}},$$

$$\tag{6.90}$$

*According to more recent critical evaluations, $H_{sep} = H_2 = 1.515$.

Figs. 6.12 and 6.13.   Example of "retarded stagnation flow" $u_\delta/u_\infty = 1 - x/L$. Laminar boundary layer.

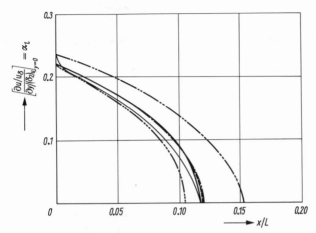

Fig. 6.12a.   Incompressible case; dimensionless tangent at the wall

$$[\partial(u/u_\delta)/\partial(y/(\delta_2)_u)]_{y=0} = \alpha_l(H)$$

as a function of x/L. Comparison of different approximations with the exact solution by Howarth [49].

—···— Method I [with eqs. (3.84), (3,86) through 3.92)], Hartree profiles
—···— Method I [with eq. (3.81)]
—— Method II (Walz, Hartree profiles)
—— Method II (Walz, Pohlhausen profiles)
— — — Method III (Geropp)
—·—· exact (Howarth)

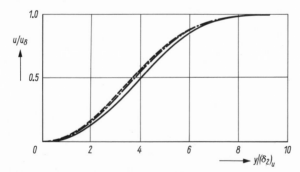

Fig. 6.12b.   Comparison of the velocity profiles obtained by different methods at the separation point.

—— Method II (Walz)
— — — Method III (Geropp)
— · — exact (Howarth)

and for Γ from eq. (4.6)

$$\Gamma(x) = Z_u \frac{u_\delta'}{u_\delta} = -\frac{a_l}{b_l + 1}\, \frac{1 - \left(1 - \dfrac{x}{L}\right)^{b_l+1}}{\left(1 - \dfrac{x}{L}\right)^{b_l+1}}. \tag{6.91}$$

The shape parameter $H(x)$ is determined from eq. (4.27) by using the Hartree velocity profiles as a basis, together with eq. (3.81), and (3.86) through (3.92) as

$$H(x) = 1{,}572 + \Gamma/1.272. \tag{6.91a}$$

The parameters are determined from eqs. (4.10) and (3.86) through (3.92) as $a_l = 0.441$, $b_l = 5.165$, for $\Gamma < 0$ ($u_\delta' < 0$).

Figure 6.12a shows a plot of the dimensionless tangent at the wall $\alpha_l(H)$ [cf. eq. (3.94)]. The separation point $\alpha_l = 0$ ($H_{\mathrm{sep}} = 1.515$, $\Gamma_{\mathrm{sep}} = -0.0681$) is found to be at $x/L = 0.1040$. Based on the Pohlhausen velocity profiles [91], eqs. (3.84), (3.85), and with $a_l = 0.470$, $b_l = 6.0$ $\Gamma_{\mathrm{sep}} = -0.1567$, the separation point is found to be farther downstream at $x/L \approx 0.153$. Both results for $\alpha_l(x/L)$ differ substantially from the exact solution.

Computational Method II, when based on the integration formulas (4.42), (4.43), yields far better agreement with the exact solution for $\alpha_l(x/L)$, and moreover, as was already shown in [140], this holds regardless of whether Hartree's or Pohlhausen's velocity profiles are used (Fig. 6.12a). This result suggests that Computational Method II is more reliable.

Computational Method III, which satisfies the momentum law, the energy law, and the compatibility condition simultaneously, produces agreement with the exact solution to within graphical accuracy. This is even true, for example, for the velocity profile at the separation point (Fig. 6.12b).

6.4.2 The laminar compressible boundary layers

Figures 6.13a and b show the influence of the Mach number and of heat transfer (parameter Θ) on the laminar separation point. The laminar separation point moves upstream with increasing Mach number. Heating of the wall produces the same effect, while cooling retards separation. For $M_\infty = 2.0$, $\Theta = 0$, the results can be compared with computations by Wrage [159] (Fig. 6.13c), which are based on finite difference methods (cf. for example [38], [116], [158]). The results obtained from Computational Method III agree also in this case, within graphical accuracy, with

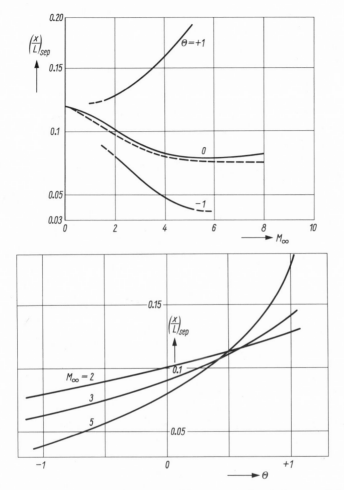

Figs. 6.13*a* and *b*. Compressible case, with heat transfer. Influence of the Mach number M_∞ and heat transfer (parameter Θ) on the laminar separation point $(x/L)_{\text{sep}}$, according to Method II.

—— $\omega = 0.7$ for $\Theta = 1; 0; -1$
– – – $\omega = 1$ for $\Theta = 0$ (Fig. 6.13*a* only)

the computations of Wrage. Computational Method II yields a separation point slightly farther downstream (at $x/L \approx 0.1$ instead of 0.092).

The results for $H(x/L)$, and thus also for $\alpha_l(x/L)$ (not shown in the graph), are independent of the Reynolds number in the case of laminar boundary layer.

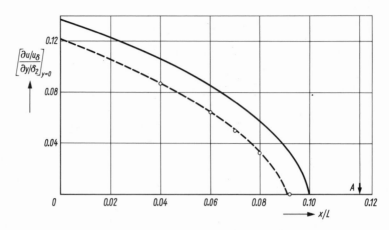

Fig. 6.13c. Graph of $[\partial(u/u_\delta)/\partial(y/\delta_2)]_{y=0}$ against x/L for $M_\infty = 2.0$; $\Theta = 0$
—— Method II (Walz)
– – – Method III (Geropp)
○   Series expansion (Wrage)
A = Separation point for $M_\infty = 0$ Howarth.

Figs. 6.14 and 6.15.   Example $u_\delta/u_\infty = 1 - x/L$. Turbulent boundary layer.

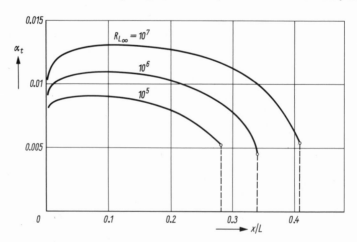

Fig. 6.14a. Incompressible case, dimensionless tangent at the wall, α_t, as a function of x/L with the Reynolds number $R_{L\infty} = \rho_\infty u_\infty \cdot L/\mu_\infty$ as a parameter.
○ = separation ($H < 1.57$).

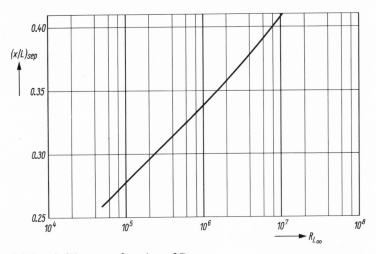

Fig. 6.14*b*. $(x/L)_{\text{sep}}$ as a function of $R_{L\infty}$.

6.4.3 The turbulent incompressible and compressible boundary layer

Figures 6.14 and 6.15, as a counterpart to Figures 6.12 and 6.13, show the results obtained with Computational Method II for turbulent boundary layer.

Figure 6.14*a* plots the influence of the Reynolds number ($R_\infty = 10^5$, 10^6, 10^7) on $\alpha_t(H)$ for $M_\infty = 0$, $\Theta = 0$. The turbulent separation point moves downstream with increasing Reynolds number (Fig. 6.14*b*). The influence of the Mach number and of heat transfer on the turbulent separation point shows, according to Figs. 6.15*a* and *b*, a tendency at high Mach numbers that differs from the laminar case (Figs. 6.13*a* and *b*): the separation point moves upstream with increasing Mach number when $M_\infty > 3$. As expected, $(x/L)_{\text{sep}}$ is in all cases far larger than for laminar boundary layer.

6.5 Tani Flow $u_\delta/u_\infty = 1 - (x/L)^n$

This flow type, illustrated in Fig. 6.16*a*, is a generalization of the flow discussed in Section 6.4, which is contained in the present case for $n = 1$. Using finite difference methods, Tani [132] and more recently Schönauer [116] have obtained exact solutions for this flow type, for laminar incompressible flows. Figure 6.16*b* shows a comparison of the results obtained with Method II for the laminar separation point $(x/L)_{\text{sep}}$, and the results by Schönauer. The agreement is good. Computational Method III has

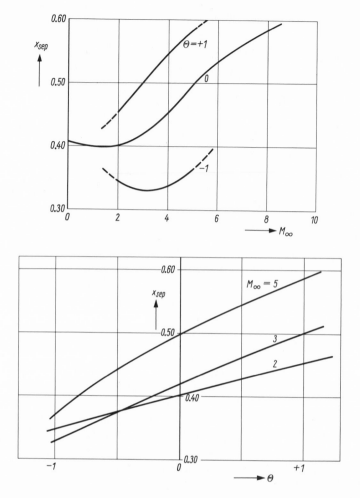

Figs. 6.15*a* and 6.15*b*.   Compressible case, with heat transfer. Influence of the Mach number M_δ and heat transfer (parameter Θ) on the turbulent separation point x_{sep} at constant Reynolds number $R_{L\infty} = \rho_\infty u_\infty L/\mu_\infty = 10^7$.

not yet been applied to this example. No computation of the turbulent boundary layer has been carried out.

6.6 *The Flow Around a Circular Cylinder*

This classical example was treated first by Blasius [4], later by Hiemenz [45], Howarth [50], Görtler [39], and others. Figures 6.17*a* and *b* show the results of different methods that were developed for this classical test case for the determination of the dimensionless tangent at the wall

Fig. 6.16. Example; Tani flow $u/u_\delta = 1 - (x/L)^n$.

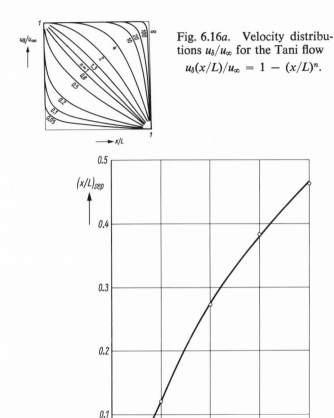

Fig. 6.16a. Velocity distributions u_δ/u_∞ for the Tani flow
$$u_\delta(x/L)/u_\infty = 1 - (x/L)^n.$$

Fig. 6.16b. Dependence of the laminar separation point upon the exponent n. Comparison with Schönauer's exact numerical solution [116].

—— Method II (Walz)
o   Finite difference method (Schönauer)

as a function of the distance x^* and for the velocity profile at the separation point. Again, the usefulness of the approximation theories manifests itself. Even Method I produces good results in this case. It should be mentioned that the computations are based on the measured data obtained by Hiemenz [45] for the pressure distribution.

Fig. 6.17.   Example: Flow about a circular cylinder.

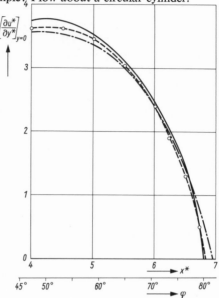

Fig. 6.17a.   Graph of the dimensionless tangent at the wall $(\partial u^*/\partial y^*)_{y=0}$ as a function of the length $x^* = x/L$; $y^* = (u_0/\nu L)^{1/2} \cdot y$; $u^* = u/u_0$, with $L = 1$ cm, $u_0 = 7.151$ cm $\cdot$ s^{-1}. Comparison of different theoretical results.

○   Finite difference method (Görtler)   ——— Method II (Walz)
$-\cdot-$ Method I (Pohlhausen) $---$ Method III (Geropp)

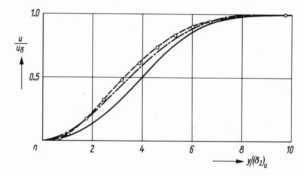

Fig. 6.17b.   Comparison of velocity profiles obtained by Methods I, II, and III at the separation point with the exact solution by Görtler [38].

○   exact (Görtler)   ——— Method II (Walz)
$-\cdot-$ Method I (Pohlhausen) $---$ Method III (Geropp)

6.7 Computation of Diffusers with Optimal Pressure Gain

The pressure gain in diffusers between two cross-sectional areas at x and $x = 0$, i.e., the "efficiency,"*

$$\eta = \frac{p(x) - p(0)}{\frac{\rho}{2}[u_\delta(0)]^2} = 1 - \left[\frac{u_\delta(x)}{u_\delta(0)}\right]^2, \tag{6.92}$$

is limited by flow separation. It should therefore be possible to determine via boundary layer computations whether a given diffuser design will work without flow separation. Boundary layer theory also answers directly the question how the cross section should increase as a function of the distance in the diffuser in order to achieve optimal pressure gain. In order for this question to have a unique answer, it is necessary to know the length of the diffuser and the "history" of the boundary layer which enters the diffuser (boundary layer thickness, shape parameter). The result of such investigations also depends strongly on whether the boundary layer is laminar or turbulent and where the transition point lies; in other words, significant influence of the Reynolds number will have to be expected. The large number of possible parameters, which may influence the result (a fact not fully recognized by the experimenter), explains why the experimental determination of optimal diffusers is difficult. Computational investigations based on today's level of achievement in boundary layer theory show better promise of success for this problem, because the different parameters that influence the result can be nicely separated from another.

Prandtl [96] already pointed out that a diffuser will have optimum efficiency, i.e., minimal losses, when the boundary layer velocity profile has a zero tangent at every point x along the diffuser wall, i.e., when the separation profile occurs everywhere. The problem of the optimal diffuser becomes thus identical with that of determining that velocity distribution $u_\delta(x)$ (with the corresponding distribution of cross-sectional areas determined by the continuity equation) which leads to the "similar solution" with the separation profile. In practice, the increases in cross-sectional area will have to be slightly less than in this theoretical optimum case for several reasons. The solution of this problem was already investigated in Sections 6.3.2, 6.3.3 for the cases of *purely laminar* and *purely*

---

*In a rigorous treatment one has to allow for the displacement effects of the boundary layer when determining the efficiency of a diffuser. On the above and other definitions of efficiency, cf. for example Sprenger [125]. There, the quantity defined in eq. (6.92) is called "pressure gain coefficient." In the sequel, we shall retain the simpler, although physically less appropriate, term "efficiency."

turbulent boundary layer, based on Computational Method II. The following discussion of the results is limited to these cases.

In the limiting case of the separation profile as a similar solution, the frictional term (the function F_2) in the momentum law (4.31) becomes zero. This simplification allows, in the case of incompressible, plane, and axisymmetric flow, for an analytic representation of $u_\delta(x)/u_\delta(0)$, while the momentum-loss thickness $\delta_2(0)$ and the Reynolds number $R_{\delta_2}(0)$ appear as parameters. Here the values $u_\delta(0)$, $\delta_2(0)$, $R_{\delta_2}(0)$ are associated with the entrance point $x = 0$ of the diffuser.

According to Fernholz [31, 32], we find for $u_\delta(x)/u_\delta(0)$ for *laminar* boundary layer

in the plane case

$$\frac{u_\delta(x)}{u_\delta(0)} = \left(1 + 0.7527\,\frac{x}{Z(0)}\right)^{-0.0904} \tag{6.93}$$

in the axisymmetric case

$$\left. \begin{array}{l} Z = \delta_2\,R_{\delta_2}, \\ (N = 1). \end{array} \right.$$

$$\frac{u_\delta(x)}{u_\delta(0)} = \left(1 + 0.6914\,\frac{x}{Z(0)}\right)^{-0.0992} \tag{6.94}$$

For *turbulent* boundary layer [with the dissipation law (3.142) through (3.144)]

in the plane case

$$\frac{u_\delta(x)}{u_\delta(0)} = \left(1 + 0.0280\,\frac{x}{\delta_2(0)\,R_{\delta_2}^N(0)}\right)^{-0.2105} \tag{6.95}$$

in the axisymmetric case

$$\left. \right\} N = 0.168.$$

$$\frac{u_\delta(x)}{u_\delta(0)} = \left(1 + 0.0245\,\frac{x}{\delta_2(0)\,R_{\delta_2}^N(0)}\right)^{-0.240} \tag{6.96}$$

In the derivation of these relations it was assumed that the separation profile (i.e., the shape parameter $H = H_{\mathrm{sep}}$) was already present at the point $x = 0$. For an arbitrary history of the boundary layer this assumption is generally not satisfied. One has rather $H(0) > H_{\mathrm{sep}}$. It is easily seen that the reduction of $H(0)$ to the value H_{sep} can be effected theoretically without losses by a velocity jump (pressure jump) (see Section 6.3.5).

Figure 6.18 shows a comparison, in the case of a plane diffuser with turbulent boundary layer, of the computed efficiency η based on eqs. (6.95) and (6.92), as a function of x, with the measurements of Stratford [129],

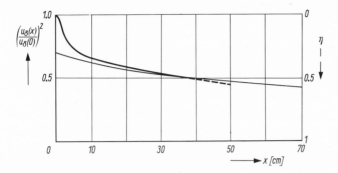

Fig. 6.18.   Dependence of the efficiency $\eta = 1 - [u_\delta(x)/u_\delta(0)]^2$ of a plane diffuser with turbulent boundary layer upon the length x of the diffuser. The velocity jump at $x = 0$ is chosen such that $H = H_{\text{sep}} = 1.57$ is reached. Comparison with measurements by Stratford [129]. Initial values at $x = 0$: $H(0) = 1.759$, $\delta_2(0) = 0.135$ cm, $R_{\delta_2}(0) = 1348$. For $x > 38$ cm the measurements by Stratford are uncertain owing to secondary flow.

—— Measurement (Stratford)

—— Computation (Fernholz)

in which the velocity profiles were realized with practically vanishing shear stress at the wall. The computations are based on the initial values given by Stratford: $\delta_2(0) = 0.135$ cm, $R_{\delta_2}(0) = 1348$, $H(0) = 1.759$; $u_\delta(0) = 15$ m/s. The measurements show that the reduction of $H(0) = 1.759$ to $H_{\text{sep}} = 1.57$ occurred in the domain between $x = 0$ and $x = 10$ cm, while the computation assumes a jump in the velocity at $x = 0$ whose magnitude is obtained from eq. (6.87) with $H(0) = H_1 = 1.759$, $H_2 = 1.57$. The agreement between computation and measurement is good, except for the initial domain. The computation shows a slightly better efficiency that obviously derives from the advantage of a velocity jump in the initial domain.

Figure 6.19 shows a comparison of the result computed from eq. (6.96) for an axisymmetric diffuser (conical diffuser) with turbulent boundary layer and the experimental results obtained by Sprenger [125]. The velocity distribution $u_\delta(x)$, which goes along with the cross-sectional distribution of the circular cone does not correspond to the optimum case with the boundary layer data used by Sprenger at $x = 0$ (this situation was not aimed for in this experiment): in the early section of the diffuser it would be desirable to have a higher velocity decrease (higher pressure increase), while farther downstream in the diffuser less of a velocity decrease would be suitable. For $x > 21$ cm, the limits of validity of Fernholz's theory [characterized by $(\delta_1/\mathfrak{R}) < 0.05$] are exceeded in the present example.

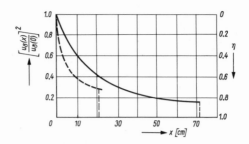

Fig. 6.19.   Circular-cone diffuser according to Sprenger [125]. Comparison of the efficiency in Sprenger's realization with the optimal solution, eq. (6.96). Initial values at $x = 0$: $H(0) = 1.612$; $R_{\delta_2}(0) = 689$; $(\delta_1/\Re)_{x=0} = 0.511\%$; $(\delta_2/\Re)_{x=0} = 0.265\%$; $\Re = 5$ cm; $u_\delta(0) = 84.76$ m · s^{-1}.

—— measurement (Sprenger)

– – – Computed (Fernholz); $H = 1.57$ const.

(At $x = 21$ cm the upper limit of the theory is reached; for $x > 21$ cm,

$$\delta_1/\Re > 0.05.)$$

How to choose the cross-sectional distribution in order to achieve optimal efficiency depends strongly upon the initial value of the quantity $\delta_2(0)\, R_{\delta_2}^{0.168}(0)$, as can be seen from eqs. (6.93) through (6.96). Figure 6.20 shows, for the case of a plane diffuser with a turbulent boundary layer which is characterized by a fixed value of the shape parameter $H = H_{\text{sep}} = 1.57$, the influence of the initial value $\delta_2(0)\, R_{\delta_2}^{0.168}(0)$ upon the efficiency and the corresponding cross-sectional distribution which follows from eq. (6.95) via the continuity condition. It is worth noting that in the limiting case of vanishing boundary layer thickness [i.e., when $\delta_2(0)\, R_{\delta_2}^{N}(0) \to 0$], the increase of the cross-sectional area may start with $(dF/dx)_{x=0} = (du_\delta/dx)_{x=0} = \infty$. Extensive development of the

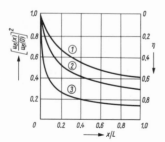

Fig. 6.20.   Plane diffuser with turbulent boundary layer. Influence of the initial momentum-loss thickness $\delta_2(0)$ upon efficiency; initial values at $x = 0$:

$H = 1.57$

$u_\delta(0) = 30.2$ m s^{-1}

$\nu = 0.151 \times 10^{-2}$ m^2 s^{-1}

$L = 1$ m

① $\delta_2(0)/L = 0.10\%$
② $\delta_2(0)/L = 0.05\%$
③ $\delta_2(0)/L = 0.01\%$

boundary layer before entering the diffuser may therefore affect the efficiency of the diffuser adversely. Boundary layer theory shows that the important parameter in eqs. (6.93) through (6.96) is the quantity $(\delta_2 R_{\delta_2}^N)_{x=0}$. For laminar boundary layer $N = 1$, for turbulent layer $N = 0.168$. Ackeret [1] and Sprenger [125] were probably the first to point out the fact that the diffuser efficiency depends to such a high degree on the history of the boundary layer prior to entering the diffuser. Relations (6.93) through (6.96) offer a simple but generally valid starting point for the design of diffusers, which (if necessary, with additional boundary layer computations) may replace hitherto used, empirical rules of limited validity.

Fernholz [32] generalized the relations (6.93) through (6.96) in such a way that the dissipation integral c_D, eq. (3.138), appears explicitly in these equations.

6.8 Examples for the Computation of Aerodynamical Properties of Airfoils

6.8.1 The pressure distribution

Figures 6.21 and 6.23 compare the computations of pressure distributions,

$$\frac{\Delta p}{q_\infty} = \frac{p - p_\infty}{q_\infty} = 1 - \left(\frac{u_\delta}{u_\infty}\right)^2; \tag{6.97}$$

$$q_\infty = \frac{\rho_\infty}{2} u_\infty^2, \tag{6.98}$$

for different airfoils based on the methods derived in Section 5.4 with experimental measurements. In addition Fig. 6.21 shows the pressure distribution (solid line) which is obtained when neglecting the boundary layer feedback [$\Delta\epsilon_0 = 0$ in eq. (5.43)]. It is seen that omitting this feedback in the computation leads to wide deviations of the results from experimental data. As long as there is no flow separation on the suction (upper) side of the profile (cf. Figs. 6.23a through f), the pressure distributions determined according to Section 5.4 agree well with the measurements. This holds also for the distribution of the lift coefficient c_a [eq. (5.47)] as a function of the angle of attack α_∞ (Figures 6.24 and 6.25). The value $dc_a/d\alpha_\infty$, which plays an important role in the mechanics of flight, when obtained from potential theory, is about 15 to 20 percent too large, as will be seen in Fig. 6.25.

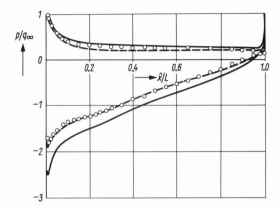

Fig. 6.21.   Pressure distribution for profile NACA 4412 according to Pinkerton [90] with and without boundary layer feedback according to Section 5.4.2 (see also Fig. 5.6). Comparison of computation and measurement at $\alpha_g = 8°$.

O   measurement
——   potential theory with $\Delta\epsilon_0 = 0$
– – –   modified potential theory according to Pinkerton [90].

Fig. 6.22.   Example of an airfoil with the maximal thickness in the rear part of the profile (laminar profile "Mustang"), which was under study at the Aerodynamische Versuchsanstalt in Göttingen.

6.8.2 The maximal lift coefficient $c_{a\,\max}$

To solve this problem, one computes first according to Section 5.4 the velocity distribution $u_\delta(x^*)$ (x^* = coordinate along the profile contour) for some c_a values up into the domain of an estimated $c_{a\,\max}$. The boundary layer computation for each value of c_a (i.e., angle of attack α_e or α_∞) starts out with a laminar boundary layer at the stagnation point, which obviously changes its position along the profile contour, depending on c_a. The limiting positions of the transition point have to be determined from Section 5.1. Experience indicates that the value $c_{a\,\max}$ can be assumed to have been reached when turbulent separation [characterized by $H_{\mathrm{sep}} \leq$ 1.57 to 1.500 at the point $x/L = 0.9$]* occurs (see Figs. 6.23f through i and 6.24).

---

*It should be noted that the rectified profile contour x^*, starting at the stagnation point $x^* = 0$, is longer in the case for $c_{a\,\max}$ than the projection x on the profile axis. Thus, $x/L = 0.9$ corresponds to a value $x^*/L > 0.9$.

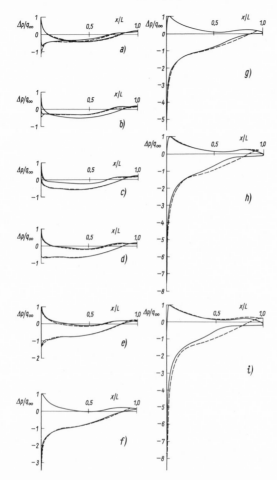

Figs. 6.23a through i.   Pressure distributions for the profile in Fig. 6.22.
——— measurement
- - - - computed according to Section 5.4 with boundary layer feedback;

$$R_L = 2.7 \times 10^6.$$

a	$\alpha_\infty = -2°$	b	$\alpha_\infty = -0.4°$	c	$\alpha_\infty = 1.1°$
d	$\alpha_\infty = 2.6°$	e	$\alpha_\infty = 4.1°$	f	$\alpha_\infty = 7.2°$
g	$\alpha_\infty = 10.9°$	h	$\alpha_\infty = 12.8°$	i	$\alpha_\infty = 13.7°$

The agreement between the computed and measured values of $c_{a\,max}$ is good, according to Figs. 6.24 and 6.25. In this comparison between measurements of $c_{a\,max}$ derived from measured pressure distributions on the one hand, and from dynamometer readings on wind-tunnel models of finite span on the other hand, one should bear in mind that differences

222

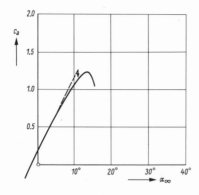

Fig. 6.24. Graph of $c_a(\alpha_\infty)$ for the profile of Fig. 6.22. Comparison of measurement and computation.

—— DVL measurement ⎫
- - - computed ⎭ $R_L = 2.7 \cdot 10^6$

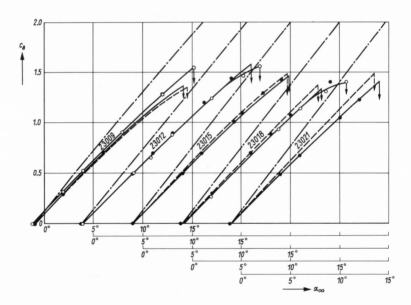

Fig. 6.25. Measured and computed values of $c_a(\alpha_\infty)$ for profile series NACA 23009, 12, 15, 18, and 21. The last two digits of the profile number indicate the thickness ratio.

Computation for $R_L = 2.6 \times 10^6$
—— taking into account the angle at the trailing edge.
- - - trailing edge angle δ_H = const. = 16°
—·— potential theory only.
Computation for $R_L \approx 2.6 \times 10^6$
○ c_a from DVL pressure distribution measurements
●   NACA measurement (weighing has been related to $\Lambda = \infty$, NACA Report No. 610.

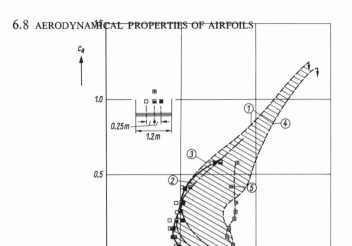

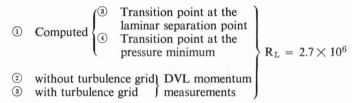

Fig. 6.26. Polars $c_a(c_{WP})$ for the "Mustang" profile. Computed for the extremal positions of the transition point according to Section 5.1. Comparison with measurements on an airfoil with endplates. The positioning of the probes is shown in the insert.

① Computed	③ Transition point at the laminar separation point	$R_L = 2.7 \times 10^6$
	④ Transition point at the pressure minimum	
② without turbulence grid	DVL momentum	
③ with turbulence grid	measurements	

are noted which are due to three-dimensional influences. The computational result applies always to the strictly two-dimensional flow. We note here also that the computation can even account for the slight increase of $c_{a\,max}$ with the Reynolds number R_L that has been observed experimentally (though not shown in Figs. 6.24 and 6.25).

6.8.3 Profile drag c_w and polar $c_a(c_w)$ of a profile

From the boundary layer quantities at the trailing edge of the profile it is possible, according to Section 5.4.3, to determine theoretically the drag coefficient $c_{WP} = C_F + C_{Dr}$, and thus also the polar $c_a(c_{WP})$. Figures 6.26 and 6.27 provide a comparison of computed and experimentally determined polars for the profile given in Fig. 6.22 and the profile NACA 23012. Of the two limiting positions of the transition point (indifference point and laminar separation point), the latter results in better agreement with experiment for the examples considered here

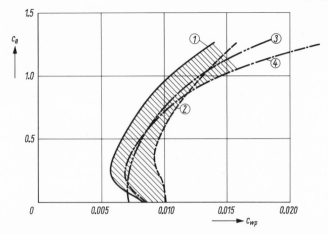

Fig. 6.27. Polars $c_a(c_{WP})$ for the profile NACA 23012. Computation based on the extremal positions of the transition point as in Section 5.1. Comparison with measurements

①   Computed with the transition point at the laminar separation point
②   Computed with the transition point at the pressure minimum
③   NACA weighing
④   DVL weighing.

(apparently owing to the low degree of turbulence of the flow). With artificial turbulence present (generated by grates in the wind tunnel), the transition point apparently moves into the vicinity of the pressure minimum ($\approx$ indifference point). These and numerous other theoretical profile investigations lead to the conclusion that the approximation theory used here for the computation of the pressure distribution and the boundary layer is capable of predicting the essential aerodynamical properties of a profile with sufficient accuracy.

The computations for the examples given in this Section 6.8 are essentially taken from unpublished reports by the author [148], [149], and [150], from the year 1943. For the laminar boundary layer, Computational Method I was used already then in the form presented here; in the domain of accelerated flow (from the stagnation point to the point of minimum pressure $\approx$ transition point) this method is practically equal in accuracy to Computational Method II. For the (larger) domain of turbulent boundary layer (approximately from the pressure minimum to the turbulent separation point), aside from the momentum integral condition, an empirical form of the integral condition for energy was used,* as given

---

*A simple form of the Gruschwitz method, reduced to quadratures, was given by A. Walz in [152] (see also [150]).

by Gruschwitz [41], in which the empirical constants were only deter-mined for incompressible flow and for a limited range of Reynolds num-bers ($0.25 \times 10^3 < R_{\delta_2} < 5.5 \times 10^3$). In this limited domain of validity, the results of Gruschwitz's method are more or less equivalent to those obtained by Computational Method II (as we shall see in Section 6.9).

6.8.4 Computation of the buffeting limit for airfoils with local supersonic domains

With modern aircraft flying at high subsonic speeds, local supersonic regions may occur on the suction side of the wing profile. These super-sonic domains are terminated by a more or less strong shock wave. Behind the shock wave the pressure rises further to the trailing edge of the profile, although very little. Although the lift coefficients c_a in this high-speed flight are much smaller than $c_{a\,max}$, as determined in Sec-tion 6.8.2, separation may occur between the starting point of the shock wave and the trailing edge of the profile. We know by experience that separation begins at the trailing edge, as in the case of low-speed flight, after $c_{a\,max}$ has been exceeded, and that the separation point moves upstream as c_a increases, but also with increasing Mach number, and it moves rather suddenly to the point of origin of the shock wave, even when these quantities increase very little. In this process lift and drag change suddenly (lift decreases, drag increases). At a slight decrease of c_a and/or the Mach number, the original flow pattern will occur again (with a certain hysteresis), and lift and drag assume their previous values just as suddenly. These changes in forces and moments represent a large mechanical load on the particular cell of the airplane as well as on the steering mechanisms, which has to be avoided.

Knowledge of the buffeting limit is therefore required for airplanes in this category. Up to now this limit, which is characterized by critical pairs of values c_a, M_∞, could only be determined (so far as the author knows) by measurements in wind tunnels and in in-flight testing. Since this phenomenon is a boundary layer effect, it should be possible in principle to determine the buffeting limit also by calculations based on Computa-tional Method II. In order to do this, the velocity distribution has to be known along the suction side of the profile together with the local super-sonic domain (the pressure side of the profile is not critical as far as lift is concerned, because of the lower pressure gradients). The result of such a boundary layer computation may be considered a success if the separa-tion point in certain domains of the values c_a, M_∞ changes its position rapidly with small variations of these quantities.

In Figs. 6.28 and 6.29 the theoretical buffeting limits obtained from a boundary layer computation [based on the dissipation law (3.142), (3.143),

Figs. 6.28 and 6.29.   Comparison of theoretical buffeting limits with those
obtained from in-flight experiments.

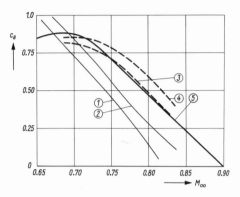

Fig. 6.28.   Airfoil NACA 65₁-110

①   Empirical criterion according to Gadd
②   Empirical criterion according to Osborne
③   Computation by Method II at $R_L = 1 \cdot 10^7$
④   Computation by Method II at $R_L = 2 \cdot 10^7$
⑤   In-flight experiment with aircraft Bell X-1 and Douglas D-558-1.

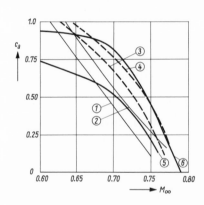

Fig. 6.29.   Airfoil NACA 65-213

①   Empirical criterion according to Gadd
②   Empirical criterion according to Osborne
③   Computation by Method II at $R_L = 1 \cdot 10^7$
④   Computation by Method II at $R_L = 2 \cdot 10^7$
⑤   Wind-tunnel results for $R_L < 2 \cdot 10^6$
⑥   In-flight experiment with aircraft Lockheed F-80A for $1.10^7 < R_L < 2 \cdot 10^7$

(3.144), (3.158)] are compared with data obtained from in-flight experiments (according to Thomas [134]*) for two different wing profiles and
three different types of aircraft. The figures also contain results on buffeting criteria obtained by Osborne and Gadd (see Pearcey and Holder [87])
as well as results from wind-tunnel measurements according to [93].

*The author is grateful to Dr. F. Thomas for permission to use computational results
contained in his habilitation thesis, (Technical University, Braunschweig). There the
reader can find details of the problems involved in this complicated problem.

Figs. 6.30 through 6.35.   Examples of retarded turbulent flows (pressure rise) for which the boundary layer quantities $H(x)$ and $\delta_2(x)$ were measured. Comparison of computations with measurements.

Measurement: ○ $H(x)$;   ● $\delta_2(x)$.

−··− Computation with eqs. (3.143) and (3.144)
−−− Computation with eqs. (3.149) and (3.150)
−·− Computation according to Gruschwitz [41]
—— Computation according to Felsch [27]

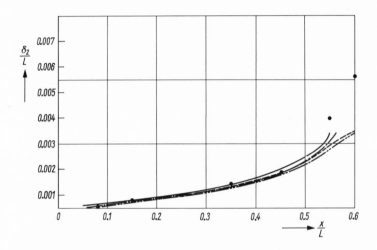

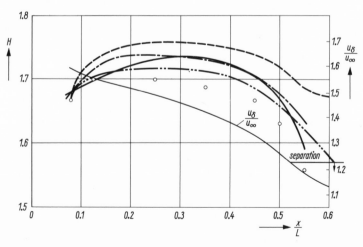

Figs. 6.30a and b.   Measurements by von Doenhoff and Tetervin [12]

A comparison shows that the boundary layer computation is, also in this case, capable of producing reliable answers and, as far as the possible influence of parameters such as the Reynolds number is concerned, apparently powerful enough to give exhaustive answers. It remains to investigate the influence of the improved dissipation law (3.156) of Felsch [27] on the results of these computations (see Section 6.9).

The following section is devoted to illustrating question of the reliability of computations of turbulent boundary layers in domains of increasing pressure by means of some characteristic examples.

6.9 Reliability of the Computation of Turbulent Boundary Layers Under Pressure Increase Based on Computational Method II

In Figures 6.30a through 6.35d a comparison is made, for different types of retarded flows $u_\delta(x)$ (pressure increase), of values of the shape parameter $H(x)$ computed according to eq. (3.75), and of the momentum loss thickness $\delta_2(x)$, with the measured values of these quantities. The three dissipation laws discussed in Sections 3.7.2 and 3.7.3 were used alternately in these computations.

We begin by noting that the momentum loss thickness $\delta_2(x)$ is practically independent of the choice of the dissipation law and agrees well with the measurements, except in one case (Fig. 6.31a) where apparently strong three-dimensional effects or influences of roughness are involved (cf. the critiques by Clauser [8] and Thompson [135]). The computed functions for the shape parameter H, however, deviate widely from one another in most cases. The values of $H(x)$ based on the dissipation law of Felsch [27] (solid curves) agree best with the measurements. One is therefore led to consider eq. (3.156) by Felsch as the most general and most reliable form of the empirical dissipation law known at the present time.

It appears important to include the results of the computational method of Gruschwitz [41], which is based on an empirical equation for the shape parameter of the velocity profile (without an explicit dissipation law) and for decades provided the only foundation for computing incompressible turbulent boundary layers. The results of this method are very useful in its domain of validity, given here as $0.25 \times 10^3 < R_{\delta_2} < 5.5 \times 10^3$ and are in general even better than those based on the c_{Di} law (3.142) through (3.144) given by Rotta [107] and Truckenbrodt [138]. The good experience with the Gruschwitz method in the examples treated in Sections 6.8.2 and 6.8.3 is therefore also understandable.

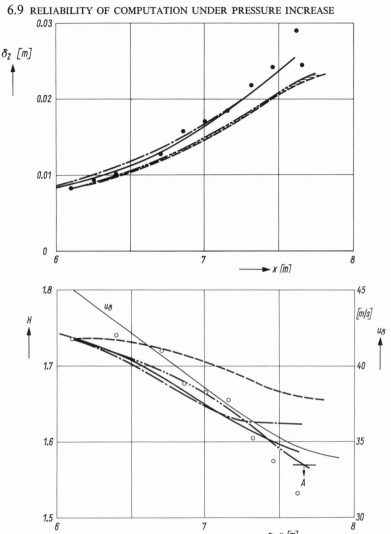

Figs. 6.31a and b.   Measurements by Schubauer and Klebanoff [118]

The energy law used in Computational Method II is, however, pre-ferred, because of its general validity (for arbitrary Reynolds numbers as well as for compressible flows with and without heat transfer), to the empirical differential equations proposed earlier, not only those by Gruschwitz but also those by von Doenhoff and Tetervin [12]. This is even more true since Felsch's work removed the uncertainties about the dissipation law.

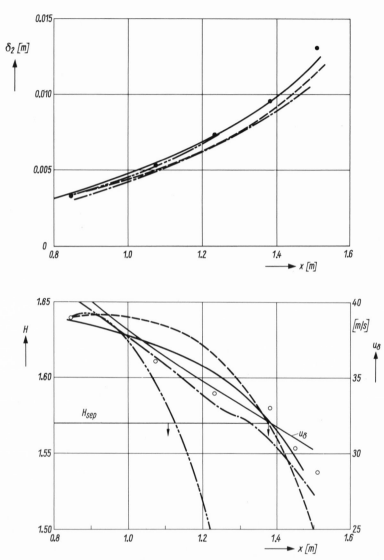

Figs. 6.32*a* and *b*.   Measurements by Newman [78]

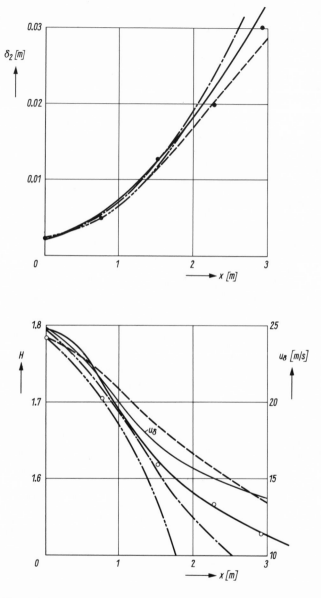

Figs. 6.33*a* and *b*.   Measurement 1 by Schubauer and Spangenberg [120]

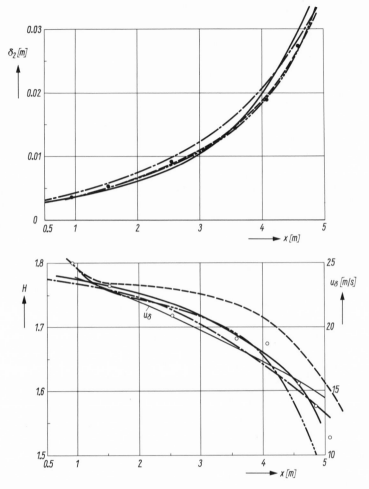

Figs. 6.34*a* and *b*.   Measurement 2 by Schubauer and Spangenberg [120]

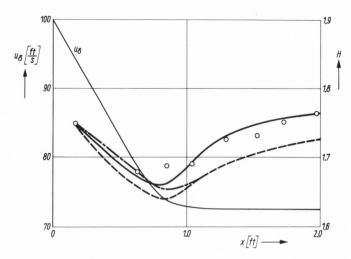

Fig. 6.35a.   Measurement 6 by Moses [76]

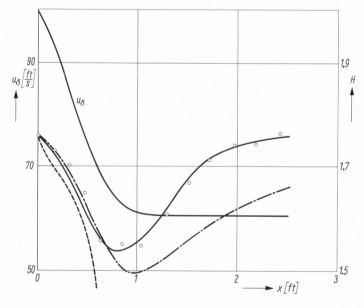

Fig. 6.35b.   Measurement 5 by Moses [76]

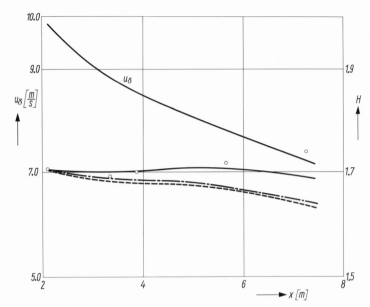

Fig. 6.35*c*.   Measurement 1 by Clauser [8]

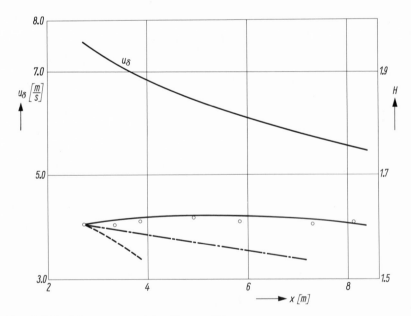

Fig. 6.35*d*.   Measurement 2 by Clauser [8]

6.10 Development of the Boundary Layer with Sinusoidally Modulated Free-stream Flow $u_\delta(x)$

6.10.1 Preliminary remarks

The reader is reminded that the approximation theory developed in the present volume for laminar and compressible boundary layers is equivalent to the exact theory when the free-stream flow $u_\delta(x)$ is of the power law type $u_\delta(x) \sim x^m$ where $m = \text{const.}$, that is, when Hartree's "similar solutions" [43] exist. The more $u_\delta(x)$ deviates from this type of flow, the greater the uncertainty in the approximate solution. A suitable test example of a flow $u_\delta(x)$ that taxes the approximation theory beyond its limits is one with a periodically, for example sinusoidally, varying velocity distribution (pressure distribution):

$$\frac{u_\delta(x)}{u_\infty} = 1 + \epsilon \sin\left(2\pi \frac{x}{\Lambda}\right). \tag{6.99}$$

Here Λ is the wave length and ϵ the relative amplitude of the stationary velocity perturbation. The velocity is scaled by the unperturbed external velocity u_∞.

This example is also of technical importance in hydrodynamics. Periodically varying velocity distributions $u_\delta(x)$ as in eq. (6.99) are found, for example, in flows along periodically undulated walls and also along plane channel walls with a periodically undulated wall on the opposite side. It is then of interest to know what influence the perturbation amplitude ϵ and the wave length Λ have on the transition phenomenon of the laminar boundary layer. If one assumes, as in Section 5.1, transition to occur at the laminar separation point (the latest possible point), the answer to this question can be found once the development of the laminar boundary layer is known up to the separation point for different values of the parameters ϵ and Λ.

The following section is devoted to the solution of this problem. Where possible, the results obtained with Approximation Methods I, II, and III will be compared with results from finite-difference methods. The influence of compressibility and heat transfer on the results will also be investigated.

6.10.2 Results

With Computational Method I, the solution for incompressible flow without heat transfer ($M_\delta = 0$, $\Theta = 0$) can be given in closed form based on the quadrature formula (4.14), provided the perturbation amplitude

$\epsilon \ll 1$, an assumption which is practically always satisfied. If for $x = 0$ we have $Z = 0$, then we find for *laminar* boundary layer:

$$\frac{Z}{\Lambda}\left(\frac{x}{\Lambda}\right) = a_l \frac{\displaystyle\int_0^{x/\Lambda} u_\delta^{b_l}\, dx/\Lambda}{u_\delta^{b_l}} \approx \frac{a_l}{2\pi} \frac{\displaystyle\int_0^{x/\Lambda} \left(1 + b_l\epsilon \sin 2\pi \frac{x}{\Lambda}\right) dx/\Lambda}{1 + b_l\epsilon \sin 2\pi \frac{x}{\Lambda}}$$

$$\approx \frac{a_l}{2\pi}\left(\frac{x}{\Lambda} - b_l\epsilon \cos 2\pi \frac{x}{\Lambda}\right)\left(1 - b_l\epsilon \sin 2\pi \frac{x}{\Lambda}\right) \tag{6.100}$$

and

$$\Gamma\left(\frac{x}{\Lambda}\right) = -\frac{Z}{\Lambda}\frac{du_\delta/d\frac{x}{\Lambda}}{u_\delta}$$

$$\approx -a_l\,\epsilon\left(\frac{x}{\Lambda} - b_l\epsilon \cos 2\pi \frac{x}{\Lambda}\right)\left[1 - (b_l + 1)\epsilon \cdot \sin 2\pi \frac{x}{\Lambda}\right], \tag{6.101}$$

$$H = 1.572 + \frac{\Gamma}{1.272}. \tag{6.102}$$

With Computational Method II the single-step formulas (4.42), (4.43), and (4.44) through (4.47) were used and the numerical solutions for different values of the parameters ϵ, Λ, M_δ, and Θ were obtained on an electronic computer.

The results from Computational Method III are taken from the doctoral thesis of Geropp [33].

Figures 6.36 through 6.38 compare the results for an example already investigated by Quick and Schröder [101], based on finite-difference methods.* The dimensionless tangent at the wall

$$\left(\frac{\partial u/u_\delta}{\partial y/(\delta_2)_u}\right)_{y=0} = \alpha_l$$

is shown as a function of the relative length x/L_{AS}.

*The velocity distribution of the potential flow associated with the examples in Figs. 6.36, 6.37 and 6.38 is given by the equations:

$$\frac{u_\delta}{u_\infty} = 1 \quad \text{for } 0 \leq \frac{x}{L_{AS}} \leq 1,$$

$$\frac{u_\delta}{u_\infty} = 1 + \epsilon \sin^2\left[2\pi \frac{L_{AS}}{\Lambda}\frac{x}{L_{AS}}\right] \quad \text{for } 1 \leq \frac{x}{L_{AS}} \leq 1 + \frac{1}{4}\frac{\Lambda}{L_{AS}},$$

$$\frac{u_\delta}{u_\infty} = 1 + \epsilon \sin\left[2\pi \frac{L_{AS}}{\Lambda}\frac{x}{L_{AS}}\right] \quad \text{for } 1 + \frac{1}{4}\frac{\Lambda}{L_{AS}} \leq \frac{x}{L_{AS}} \to \infty.$$

In this problem the result thus depends also on the starting length L_{AS}.

Figs. 6.36 through 6.38.   Example of sinusoidally modulated external flow according to Quick and Schröder [101], in incompressible flow ($M_\infty = 0$; $\Theta = 0$), for different values of the perturbation amplitude ϵ and the starting length L_{AS}. Comparison of the results of the different computational methods for laminar boundary layers.

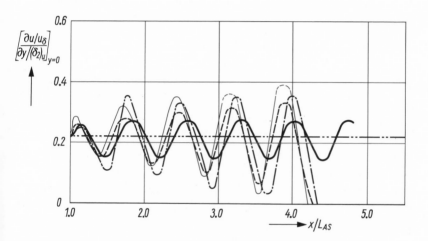

Fig 6.36. $\epsilon = 0.01$; $\Lambda/L_{AS} = 0.72$; dimensionless tangent at the wall $[\partial(u/u_\delta)/\partial(y/\delta_{2u})]_{y=0}$ as a function of the nondimensional length x/L_{AS}.

—— Method I (Pohlhausen)
—— Method II (Walz)
– – – Method III (Geropp)
– · – Finite difference method (Quick and Schröder)
– ·· – $\epsilon = 0$ (Flat-plate boundary layer).

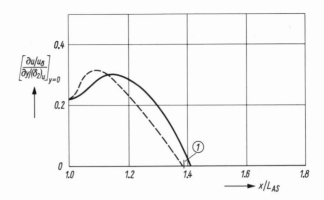

Fig. 6.37. Perturbation amplitude $\epsilon = 0.025$; $\Lambda/L_{AS} = 0.72$. Dimensionless tangent at the wall $[\partial(u/u_\delta)/\partial(y/\delta_{2u})]_{y=0}$ as a function of the non-dimensional length x/L_{AS}

—— Method II (Walz)
– – – Method III (Geropp)
① Separation, as determined by finite difference method (Quick and Schröder)

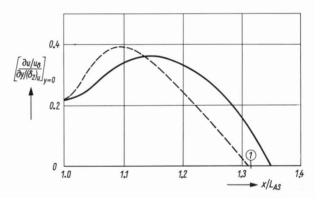

Fig. 6.38. Perturbation amplitude $\epsilon = 0.05$; $\Lambda/L_{AS} = 0.72$. Demensionless tangent at the wall $[\partial(u/u_\delta)/\partial(y/\delta_{2u})]_{y=0}$ as a function of the dimensionless length x/L_{AS}

—— Method II (Walz)
– – – Method III (Geropp)
① Separation, as determined by finite-difference method (Quick and Schröder)

For a perturbation amplitude $\epsilon = 0.01$ and a fixed ratio of the "starting length" L_{AS} (during which the flat-plate boundary layer is unperturbed) to the wave length Λ of the perturbation, $\Lambda/L_{AS} = 0.72$, Computational Method III puts the separation point at about $x/L_{AS} = 4.3$, in good agreement with the results of Quick and Schröder. Computational Method I gives surprisingly useful results. However, the separation point can be determined by this method only if one allows, in domains of decreasing pressures, velocity profiles to occur with physically meaningless velocities that exceed the free-stream velocity (partly dashed in Fig. 6.36). Computational Method II does not indicate separation at all. With enlarged perturbation amplitudes $\epsilon = 0.025$ and 0.05, which according to Figs. 6.37 and 6.38 lead to flow separation already after the first pressure wave, both Computational Methods I and II give useful results. For periodically varying pressure, or more generally when the pressure gradient frequently changes its algebraic sign in the direction of flow, we suspect that probably only a finite-difference method will be applicable. However, Computational Method III turns out to be almost as powerful as the finite-difference methods. This conclusion is supported also by Fig. 6.39, which shows a comparison of the laminar velocity profiles at the separation point $x/L_{AS} = 4.3$ for the example of Fig. 6.36.

In this example of a sinusoidally modulated free-stream flow $u_\delta(x)$ it is also interesting to determine the influence of compressibility and heat transfer. Only the results for Computational Methods II and III will be compared. Theoretically exact comparison values are not known in the numerical example treated here.*

The starting length L_{AS} is assumed to be 0 and, for all computations, the perturbation amplitude is $\epsilon = 0.005$. The results of the computations are illustrated in Figs. 6.40 through 6.43. One finds, based on Computational Method III, that the influence of increasing Mach number M_∞ is such that the separation point x/Λ moves from about 13.5 at $M_\infty = 0$ (Fig. 6.40) to $x/\Lambda \approx 4.3$ at $M_\infty = 3.19$ (Fig. 6.41). If at this particular Mach number the wall is cooled such that $T_W = T_\delta$, i.e., $\Theta = 1$, then the separation point moves downstream to about the point $x/\Lambda = 13.5$ (Fig. 6.42) which was the separation point for $M_\infty = 0$.

Computational Method II does not indicate separation in the cases investigated. The trends of the influences of $M_\infty^*(M_\infty)$ and Θ upon the shape parameter and the shear stress are, however, the same. A comparison of the velocity profiles computed with Computational Methods II and

---

*After having finished the computations, the author became aware of a paper by T. Fannelöp and I. Flügge-Lotz [26], where similar examples are treated by a finite-difference method.

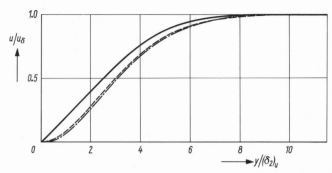

Fig. 6.39.   Velocity profiles at the point $x/L_{AS} \approx 4.3$ where, according to
Fig. 6.36, Method III yields vanishing shear stress (roughly in agreement
with Quick and Schröder's result).

$\epsilon = 0.01$ $M_\infty = 0$; $\Theta = 0$; $\Lambda/L_{AS} = 0.72$; $x/L_{AS} \approx 4.3$
—— Method II (Walz)
– – – Method II (Geropp)
– · – Finite difference method (Quick and Schröder)

Figs. 6.40 through 6.43.   Example of sinusoidally modulated free-stream flow
according to eq. (6.99) without starting length. Perturbation amplitude
$\epsilon = 0.005$. Comparison of results from different computational methods for
incompressible and compressible laminar boundary layer, without and with
heat transfer.

—— Method II (Walz);   – – – Method III (Geropp); A = Separation point

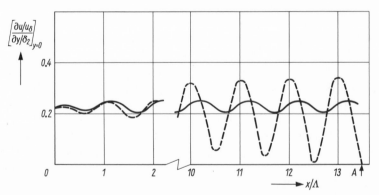

Fig. 6.40.   Incompressible flow $M_\infty = 0$; $\Theta = 0$; $\epsilon = 0.005$. Dimensionless tangent at the wall $[\partial(u/u_\delta)/\partial(y/\delta_{2u})]_{y=0}$ as a function of the dimensionless length x/Λ; Λ = wavelength of the perturbation.

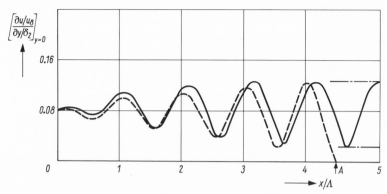

Fig. 6.41. Compressible flow $M_\infty^* = 2.0$; $M_\infty = 3.19$; $\Theta = 0$; $\omega = 1$ (heat-insulated wall)

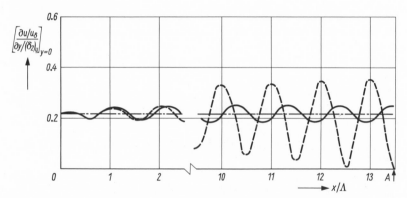

Fig. 6.42. Compressible flow $M_\infty^* = 2.0$; $M_\infty = 3.19$; $\Theta = 1$; $\omega = 1$ (considerable cooling of the wall, such that $T_w = T_\delta$)

III, Fig. 6.43, at the point $x/\Lambda \approx 13.5$ of the example of Fig. 6.42 ($M_\infty = 3.19$, $\Theta = 1$), is very informative: the velocity profile obtained from Computational Method II agrees completely, throughout the entire computation, with the more precise result of Method III. Because of the fact that the Hartree velocity profiles contain only one parameter, it is impossible with Method II to satisfy the compatibility condition (3.62) at the wall which requires that the shear stress vanishes at the point $x/\Lambda = 13.5$.

In the domain of validity of Computational Method II this specific question was also investigated: At which perturbation amplitude does separation occur after one single pressure wave, as illustrated in Figs. 6.37

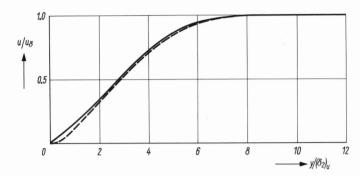

Fig. 6.43. Velocity profiles at the point $x/\Lambda \approx 13.5$ of the example in Fig. 6.42, where the shear stress vanishes according to Method III. Method II (Walz) yields a velocity profile which differs from that of the more precise solution by Geropp only in the immediate vicinity of the wall.

and 6.38? The assumed starting length was $L_{AS} = 0$. Figure 6.44 shows the results of this computation in the form of critical curves $\epsilon_{crit}(M_\infty^*, \Theta)$. At $M_\infty = M_\infty^* = 0$, the critical amplitude is found to be $\epsilon_{crit} = 0.0213$. It decreases with increasing Mach number but increases if the wall is sufficiently cooled, $\Theta > 0.5$. If one assumes that the laminar separation point occurs at the latest possible point of transition of the laminar boundary layer into the turbulent state, following the ideas developed in Section 5.1, the computations show that this transition is favored as the Mach number increases but is retarded or even completely prevented if the wall is cooled.

The results obtained for purely stationary perturbations show parallels to the results of classical stability theory which are not immediately physically evident, since classical stability theory is based on trial solutions

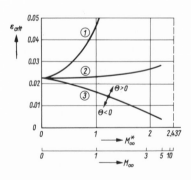

Fig. 6.44. Critical amplitude ϵ_{crit} of a stationary flow according to eq. (6.99) which leads to laminar separation within the first pressure wave, plotted as a function of the Mach number $M_\infty^*(M_\infty)$ and the heat transfer parameter Θ.

① $\Theta = 1$ $(T_w = T_\delta)$
② $\Theta = 0.5$
③ $\Theta = 0$
    $\Theta < 0$ (heating)
    $\Theta > 0$ (cooling)

of a nonstationary nature for the perturbations. An increase of the perturbation amplitude *with time* at a fixed point x in classical stability theory corresponds to a greater danger of separation (= the danger of transition) at constant (stationary) perturbation amplitude with respect to *the length x*. Increasing the Mach number with insulated or heated wall reduces the critical Reynolds number found by stability theory $(R_{\delta_2})_{\text{crit}}$, while cooling increases it to the point where for sufficiently strong cooling to arbitrarily large values absolute stability can be reached (see van Driest [15, 16] and Dunn and Lin [18]. The parallel to this result of stability theory in the treatment of stationary perturbations is the total disappearance of laminar separation when the wall is sufficiently cooled ($\Theta > 1$).

6.11 Development of the Boundary Layer with Turbulent Free-stream Flow

This problem is generally posed with hydraulic machinery whose boundary layers at the walls are under the influence of a strong turbulent main flow, but also in wind or water tunnels with a high degree of turbulence. The difference between this and the problem posed in the preceding section consists in the fact that the perturbation is nonstationary and that a whole spectrum of perturbation wave lengths (frequencies) is in action. This problem differs in concept from the usual stability theory by the fact that one no longer asks when amplification of the perturbation amplitude occurs with time. The changes of the perturbation quantities are already given here as functions of space and time and now the question is how the laminar boundary layer develops under these conditions.

Since only the theory of stationary boundary layer phenomena is treated within the framework of this volume, we shall have to defer the treatment of this problem to a later publication. However, it seems interesting to consider experimental results pertaining to it, such as measurements concerning the influence of the degree of turbulence

$$\epsilon' = \frac{\sqrt{\bar{u}^2}}{u_\delta} \tag{6.103}$$

on the position of the transition point. [The bar indicates the time averages; cf. for example, eq. (1.17).] If the perturbation velocity $\bar{u}$ varies sinusoidally, that is

$$\frac{\bar{u}}{u_\delta} = \epsilon \sin\left(2\pi \frac{x}{\Lambda}\right), \tag{6.104}$$

then the following relation exists between the degree of turbulence, eq. (6.103), and the perturbation amplitude ϵ:

$$\epsilon' = \epsilon \sqrt{\frac{1}{2\pi} \int_0^1 \sin^2 \left(2\pi \frac{x}{\Lambda} \right) d \left(2\pi \frac{x}{\Lambda} \right)} = \epsilon \sqrt{\frac{1}{2}} = 0.707\epsilon \qquad (6.105)$$

(the average of $\sqrt{\bar{u}^2}/u_\delta$ is taken over one complete cycle Λ). ϵ and ϵ' are therefore of the same order of magnitude.

In a comprehensive summary of experimental results by Granville [40] (Fig. 6.45) one finds that the experimentally observed transition point, characterized by the Reynolds number $(R_{\delta_2})_U$, moves upstream with increasing degree of turbulence (to smaller values of R_{δ_2}) and practically coincides with the indifference point of stability theory, which is characterized by $(R_{\delta_2})_{\text{crit}}$ as soon as the degree of turbulence $\epsilon' \geq 0.024$ or when $\epsilon \geq 0.017$. For the *stationary* perturbation given by eq. (6.99), laminar separation was obtained after a length $x < \Lambda$ (practically independent of the relative wave length Λ/L), as soon as the perturbation amplitude $\epsilon > 0021$. It is surprising that the numerical values of ϵ and ϵ' agree almost completely in these two cases, although they are actually not physically comparable. Presumably transition occurs at the same point as laminar separation even when the laminar boundary layer is under the influence of a nonstationary perturbation in the free-stream flow (cf. the experimental measurements by L. Dryden in Fig. 6.45, where transition to the turbulent state is observed before the indifference point). Boundary layer computations concerning this nonstationary type of perturbation in the free-stream flow are planned for the purpose of shedding additional light on this problem.

The question of *influencing turbulent boundary layers by the turbulence of the free-stream flow* is of far-reaching practical importance. We shall confine ourselves to pointing out a few references concerned with clarifying this problem both theoretically and experimentally [58].

6.12 Examples for an Improved Computation of Heat Transfer According to Section 5.3*

6.12.1 Preliminary remarks

The computation of heat transfer according to relation (2.175) is so simple that no illustration by example is necessary: once the local coeffi-

*The theoretical derivation of the solutions and their numerical treatment for the examples in this section are due to Dr. M. Mayer (cf. also [71]). The author acknowledges the use of these results.

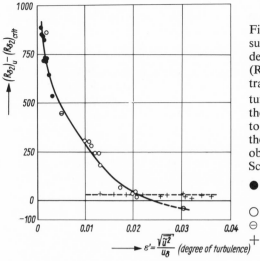

Fig. 6.45. Experimental results concerning the dependence of the Reynolds number $(R_{\delta_2})_u$ for laminar-turbulent transition upon the degree of turbulence $\epsilon' = \sqrt{\overline{u^2}}/u_\delta$ of the free-stream flow, according to Granville [40]. $(R_{\delta_2})_{crit}$ is the critical Reynolds number obtained from Tollmien-Schlichting's stability theory.

●   Schubauer,
    Skramstadt } flat plate
○   Hall-Hislop
⊖   Dryden
+   Schubauer, elliptic cylinder, with pressure rise.

cient of friction $c_f(x)$ is known from one of the Computational Methods I through III, the parameter Θ of heat transfer is known in any problem and thus also the local heat flux $q_w(x)$. It is however necessary that the assumptions made in deriving eq. (2.175), i.e., $\Pr \approx 1$, $dT_w/dx = 0$, $dp/dx = 0$, be approximately satisfied.

So long as only the flow boundary layer is of interest, as for example the coefficient of friction, the separation behavior, or the displacement action of the boundary layer, eq. (2.175), as already mentioned, can still be used as a good approximation to determine the integral expressions (3.34) through (3.36), even if these assumptions are badly violated, because approximations for the temperature and density distribution $T/T_\delta = \rho_\delta/\rho$ are sufficient to determine the averages.

For computations of heat transfer, where the tangent at the wall $(\partial T/\partial y)_{y=0}$ enters with its full weight, the function $T(x, y)$ has to be well known, particularly in the vicinity of the wall. The approximation theory developed in Section 5.3 for the computation of $T(x, y)$, or better $T(u, x)$, yields acceptable accuracy in the computation of heat transfer, even in strong temperature and pressure gradients in the direction of flow, both for laminar and turbulent boundary layers. This will be shown by some characteristic examples for which some exact comparisons are available. The case of arbitrary Prandtl number can be dealt with in principle by applying the modified Reynolds analogy of Section 5.3, but this has not been tried as yet. The examples to be discussed are therefore confined to the case where $\Pr = 1$ or $\Pr \approx 1$.

6.12.2 The flat plate with locally varying wall temperature, constant material properties (incompressible), Pr $= 1$

6.12.2.1 The laminar boundary layer. This example has been treated by Schlichting [114]. The temperature difference $T_w - T_\delta$, i.e., the parameter b used in Section 5.3, is assumed to be so small that ρ/ρ_δ is practically equal to unity. Under this assumption the flow boundary layer in the present example is identical with the flat plate boundary layer of Blasius [4], whose essential data are given in Section 6.2.2 (in complete agreement with the exact solution). In order to solve eq. (5.17), no iteration is necessary here. Even the correction function $K(x)$ can be obtained explicitly if one makes the (not necessarily required) assumption that for $x = 0$ also $K = 0$:

$$K\left(\frac{x}{L}\right) = \frac{1}{0.572}\left(\frac{x}{L}\right)^{-1.374} \cdot \int_0^{x/L} \frac{db}{dx}\left(\frac{x}{L}\right)^{1.374} d(x/L). \qquad (6.106)$$

Any arbitrary function may now be given for $b(x/L)$ and thus, $db/d(x/L)$ will also be known [Schlichting makes the number of necessary coefficient functions dependent on $b(x/L)$].

The specific examples investigated by Schlichting assume the function $T_w(x)/T_\delta$ to be linear or quadratic in its dependence on x. For

$$\frac{T_w}{T_\delta} = 1 - b = 1 + \left(\frac{T_w(0)}{T_\delta} - 1\right)\left(1 - 2\frac{x}{L}\right);$$

$$[T_w(0) = (T_w)_{x=0}] \qquad (6.107)$$

or

$$\frac{T_w}{T_\delta} = 1 + \left(\frac{T_w(1)}{T_\delta} - 1\right)\left[2\frac{x}{L} - \left(\frac{x}{L}\right)^2\right], \qquad (6.108)$$

one obtains with eq. (6.106) for the local heat flux $q_w(x)$ the solution

$$\frac{q_w(x/L)}{\lambda_w \dfrac{T_w(0) - T_\delta}{L}\sqrt{R_L}} = 0.332\left(\frac{x}{L}\right)^{-1/2} - 1.153\left(\frac{x}{L}\right)^{1/2} \qquad (6.109)$$

if one uses (6.107) and

$$\frac{q_w(x/L)}{\lambda_w \dfrac{T_w(1) - T_\delta}{L}\sqrt{R_L}} = 1.153\left(\frac{x}{L}\right)^{1/2} - 0.676\left(\frac{x}{L}\right)^{3/2} \qquad (6.110)$$

if one assumes (6.108). The solutions given by Schlichting for these two examples are:

$$\frac{q_w(x/L)}{\lambda_w \dfrac{T_w(0) - T_\delta}{L} \sqrt{R_L}} = 0.332 \left(\frac{x}{L}\right)^{-1/2} - 1.06 \left(\frac{x}{L}\right)^{1/2} \tag{6.111}$$

and

$$\frac{q_w(x/L)}{\lambda_w \dfrac{T_w(1) - T_\delta}{L} \sqrt{R_L}} = 1.06 \left(\frac{x}{L}\right)^{1/2} - 0.635 \left(\frac{x}{L}\right)^{3/2} \tag{6.112}$$

Figures 6.46 and 6.47 show that the solution developed here differs but little from Schlichting's, but that the error is considerable compared to the computations according to eq. (2.175). The heat flux $q_w(x)$ in the case (6.107), Fig. 6.46, does not vanish at $x/L = 0.5$, where the temperature

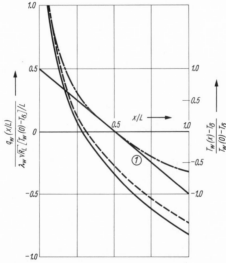

Fig. 6.46. Graph of the dimensionless heat flux. Flat plate, laminar flow, constant material properties and $Pr = 1$. Linear relation for the temperature at the wall according to eq. (6.107)

—— Theory with $\dfrac{T}{T_\delta}\left[\dfrac{u}{u_\delta}, K(x)\right]$

– · – Theory with $\dfrac{T}{T_\delta}\left[\dfrac{u}{u_\delta}\right]$; $K = 0$

– – – Schlichting's theory [114]

① Temperature, T_w, at the wall

difference between the wall and the free-stream flow vanishes, but further upstream at $x/L = 0.288$ (0.313 in Schlichting's paper).

For the example shown in Fig. 6.46 we have plotted in Fig. 6.48 also the temperature profiles $T/T_\delta - 1$ at the points $x/L = 0.288$ and 0.5 (with $T_w(0)/T_\delta = 1.1$). One recognizes that at the point $x/L = 0.5$, despite of $T_w = T_\delta$, the tangent at the wall $(\partial T/\partial y)_{y=0} \neq 0$, which means that the heat flux q_w does not vanish at that point. On the other hand, at the point $x/L = 0.288$, despite of $T_w \neq T_\delta$, the tangent at the wall $(\partial T/\partial y)_{y=0} = 0$ and thus $q_w = 0$. The local heat flux q_w together with $K(x)$ can by the way be computed in every case from the simple relation (5.23).

6.12.2.2 The turbulent boundary layer. The approximation theory in Section 5.3 allows, as mentioned earlier, the treatment of any turbulent boundary layer problem without additional complications. Empirical relations already used in Section 6.2.3, are brought into play here, in particular one for the local drag coefficient c_f. For constant material properties it is here also often possible to integrate eq. (5.17) to obtain $K(x)$ explicitly.

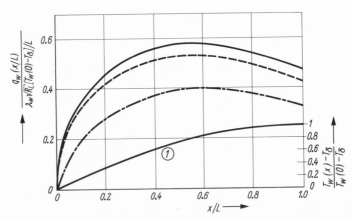

Fig. 6.47. Graph of the dimensionless heat flux. Flat plate, with the wall temperature determined by a second-degree polynomial according to eq. (6.108). Laminar boundary layer, constant material properties and Pr $= 1$.

—— Theory with $\dfrac{T}{T_\delta}\left[\dfrac{u}{u_\delta}\;;\;K(x)\right]$

— · — Theory with $\dfrac{T}{T_\delta}\left[\dfrac{u}{u_\delta}\right]\;;\;K = 0$

— — — Theory by Schlichting [114]

① Temperature, T_w, at the wall.

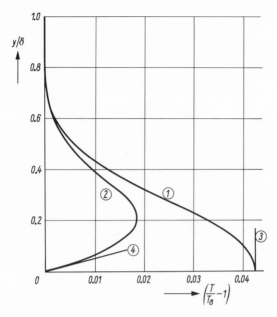

Fig. 6.48.   Temperature profiles $T/T_\delta - 1$ for the example of Fig. 6.46 at the points $x/L = 0.288$ and 0.50 for $T_w(0)/T_\delta = 1.1$

① $x/L = 0.288$ ② $x/L = 0.5$ ③ $(\partial T/\partial y)_w = 0$ ④ $(\partial T/\partial y)_w \neq 0$

For the linear distribution of the wall temperature, eq. (6.107), one obtains for the heat flux $q_w(x/L)$, in complete analogy to eq. (6.109)

$$\frac{q_w(x/L)}{\lambda_w \dfrac{T_w(0) - T_\delta}{L} R_L^{0.846}} = 0.0148 \left(\frac{x}{L}\right)^{-0.154}$$

$$- \frac{0.0493\, H}{1.668\, H - 1} \left(\frac{x}{L}\right)^{0.846} \qquad (6.113)$$

(Fig. 6.49). The heat flux q_w vanishes here at $x/L \approx 0.33$, i.e., approximately at the same point as in the laminar boundary layer. Since the shape parameter H depends on the Reynolds number [cf. eq. (6.25)], a slight dependence of the result on H is observed here:

H	1.700	1.750	1.800	1.850	laminar $H = 1.572$
$(x/L)_{q=0}$	0.324	0.329	0.333	0.338	0.288 (0.313 according to Schlichting)

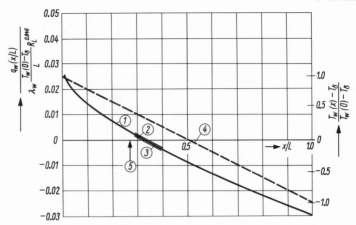

Fig. 6.49. Graph of the dimensionless heat flux. Flat plate with linearly changing wall temperature $T_w(x/L)$, according to eq. (6.107). Turbulent boundary layer from $x/L = 0$ to 1, constant material properties, and Pr = 1.

① $H = 1.75$
② $H = 1.85$
③ $H = 1.70$
④ Graph of wall temperature
⑤ For comparison, the value for x/L when $q_w = 0$ in the laminar case is indicated (see Fig. 6.46).

6.12.3 Flows with pressure gradients, of the type $u_\delta \sim x^m$, but with constant wall temperature, laminar boundary layer, Pr = 1

This example is intended to clarify the important question how the pressure gradient in the direction of flow influences the heat transfer when the wall temperature is held constant. In order to simplify the problem, which was already treated more generally in compressible flow by Li and Nagamatsu [66] (see also Figs. 2.7 and 2.8), we assume here constant material properties. In this case, the exact similar solutions given by Hartree [43] are valid, and eq. (5.17) can be integrated in explicit form. One finds

$$\frac{K(x)}{b(x)} = -m(H) \frac{1 + H_{12}(H)}{\dfrac{H}{2} + m(H)\left[1 + \dfrac{H}{2} + H_{12}(H)\right]} = f(H). \quad (6.114)$$

Figure 6.50 shows K/b as a function of $H(m)$. For the case of the flat plate, with $m = 0$, it follows from (6.114), as required, that $K = 0$. In accelerated flow, $m > 0$, K/b is negative, and the heat flux, according to eq. (5.23)

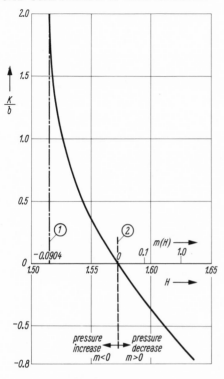

Fig. 6.50. Influence of the pressure gradient for a flow of the type $u_\delta \sim x^m$ on the heat transfer, for T_w = const., laminar boundary layer, constant material properties, and Pr = 1.

① H_{sep} = 1.515,
m_{sep} = −0.0904
② H = 1.572,
m = 0 (flat plate)

is thus smaller than would be expected from the classical Reynolds analogy, eq. (2.175). In retarded flow (pressure increase), $m < 0$, the heat transfer is larger than expected from eq. (2.175).

When large (positive or negative) pressure gradients dp/dx are present, the effect on the heat transfer is therefore considerable. A verification of this somewhat surprising and important result, either by finite difference methods or by experiment, is still outstanding. A closer inspection of Fig. 2.8, which illustrates the results of the exact similar solutions by Li and Nagamatsu, shows that varying the parameter $\beta^*(m)$ [which is responsible for the pressure gradient, cf. eq. (6.114)] between $\beta^* = 0$ and $\beta^* = 1$ is accompanied by changes of $(\partial h/\partial \eta)_{\eta=0} = c_p(\partial T/\partial y)_{y=0} \sim q_w$ of up to about 30 percent, although the function $h(\eta)$ itself, and therefore also $T(\eta)$, is hardly effected by changes of β^* (cf. Schlichting [111]). The result of this investigation concerning the influence of dp/dx on the heat transfer, which was carried out here for constant material properties, is therefore not in contradiction to the exact results of Li and Nagamatsu in compressible flow.

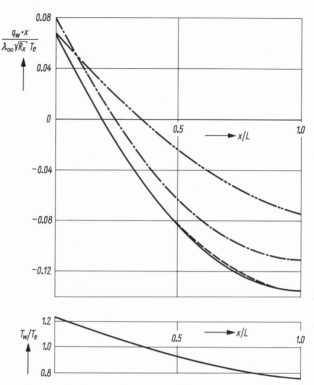

Fig. 6.51.   Comparison of different computations for the heat flux at a flat
plate in compressible laminar flow with $M_\infty = 3.0$, $Pr = 0.72$, $\mu \sim T$, and
wall temperature $T_w(x)/T_e = 1.25 - 0.83x/L + 0.33(x/L)^2$.
—— Walz [151] (approximation)
– – – Chapman and Rubesin [7] (exact)
– · – Morris and Smith [75] (approximation)
– ·· – $T_w = $ const. $(K = 0)$.

*6.12.4   The flat plate with an arbitrary distribution of the wall temperature
in a compressible laminar boundary layer*

This problem was solved exactly by Chapman and Rubesin [7] with the
aid of a suitable coordinate transformation. The numerical treatment of
the solution, however, requires an analytical approximation of $T_w(x)$ by
a polynomial. The above-mentioned paper therefore contains solutions
only for special problems. These solutions may be used as test examples
to check the accuracy of approximation theories.

In Fig. 6.51 the result of the approximation theory of Section 5.3 is
compared with the exact solution according to [7] and one other approxi-

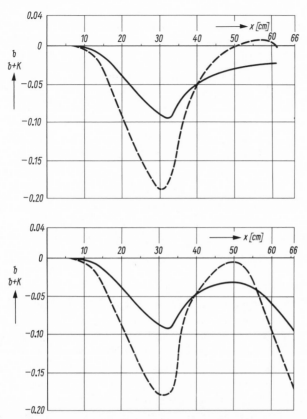

Figs. 6.52 and 6.53. Examples for distributions of the wall temperature $b(x) = 1 - T_w(x)/T_\delta$ along a cylinder in axial flow, according to Eckert et al. [20]; the function $b + K$ is computed for laminar boundary layer, constant material properties, and Pr $= 1$.

—— b
--- $b + K$

mate solution given by Morris and Smith [75]. The computation was based on the temperature distribution $T_w(x)$ at the wall

$$\frac{T_w(x)}{T_e} = 1.25 - 0.83 \frac{x}{L} + 0.33 \left(\frac{x}{L}\right)^2 \tag{6.115}$$

at $M_\infty = 3.00$ and Pr $= 0.72$ for air, and the viscosity law $\mu \sim T^\omega$, with $\omega = 1$. The result of the approximation theory of Section 5.3 agrees well with the exact solution and certainly better than the results of Morris and Smith. It is worth noting that for the approximation solution in Section 5.3, the function $T_w(x)$ may be arbitrary, and it is not required that it can be approximated by a polynomial in x/L. $T_w(x)$ may therefore be even discontinuous.

Figure 6.51 also contains the local heat flux $q_w(x)$ for $K = 0$, i.e., for $T_w = \text{const}$. The large error, which results from neglecting the temperature gradient dT_w/dx, can be immediately recognized.

6.12.5 A cylinder in axisymmetric flow with locally varying wall temperature in laminar and turbulent boundary layer, constant material properties, $Pr = 1$

For this case of axisymmetric flow without pressure gradients, but with temperature gradients along the wall, there are experimental measurements of temperature profiles in laminar and turbulent boundary layer in the literature (Eckert, Eichhorn and Eddy [20]), Figs. 6.52 and 6.53 show, for laminar boundary layer, the experimentally determined distribution of the parameter $b(x) = (T_\delta - T_w(x))/T_\delta$, and the function $b + K$, where $K(x)$ was computed according to eq. (5.17). According to eq. (5.23), the local heat flux is proportional to $b + K$. The difference between $b(x)$ and $b(x) + K(x)$ indicates that a computation with $K = 0$ would introduce considerable errors. Figures 6.54 and 6.55 illustrate what improvement can be achieved in the temperature profile $(T - T_\delta)$ when observing the correction function $K(x)$ in eq. (5.16).

Figures 6.56 through 6.59 show the result of a corresponding computation for the experimental examples in a turbulent boundary layer. The tendency of improvement is also apparent here when observing $K(x)$ in the calculations; however, the agreement with the measured temperature profiles at larger distances from the wall is not as good as for laminar boundary layer. In judging this comparison, one should bear in mind, however, that the temperature distribution in the vicinity of the wall [more precisely, the tangent at the wall $(\partial T/\partial y)_{y=0}$] determines the local heat flux.

Figs. 6.54 and 6.55.   Computed temperature profiles $T(y) - T_\delta$ for the examples of Figs. 6.52 and 6.53 for laminar boundary layer. Comparison with measurements by Eckert et al [20].

●   measurements by Eckert et al.
—— temperature function of eq. (5.16) with $K(x)$
– – – temperature function of eq. (5.16) with $K = 0$

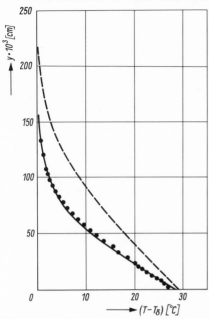

Fig. 6.54. $(T - T_\delta)$ corresponding to Fig. 6.52 at $x = 32$ cm

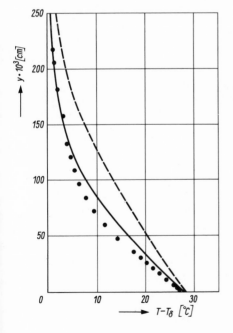

Fig. 6.55. $(T - T_\delta)$ corresponding to Fig. 6.53 at $x = 66$ cm

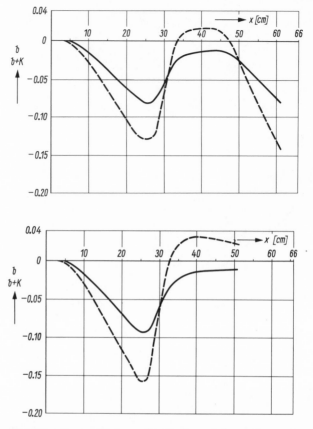

Figs. 6.56 and 6.57. Examples for a distribution of the wall temperature $b(x) = 1 - T_w(x)/T_\delta$ along a cylinder in axial flow, according to Eckert et al. [20], together with computed curves for the function $b + K$ for turbulent boundary layer.

—— b
--- $b + K$

Figs. 6.58 and 6.59. Computed temperature profiles, $T(y) - T_\delta$, corresponding to Figs. 6.56 and 6.57; turbulent boundary layer. Comparison with measurements by Eckert et al. [20].

●   measurement by Eckert et al
—— temperature function, eq. (5.16), with $K(x)$
--- temperature function, eq. (5.16), with $K = 0$.

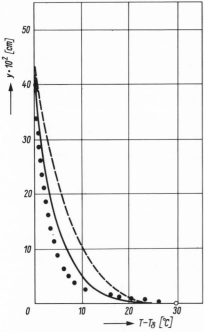

Fig. 6.58. $(T - T_\delta)$ corresponding to Fig. 6.56 at $x = 61$ cm

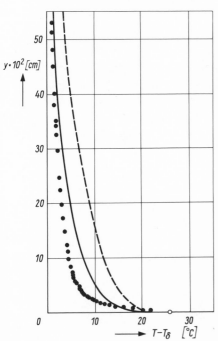

Fig. 6.59. $(T - T_\delta)$ corresponding to Fig. 6.57 at $x = 25.4$ cm

Appendix I

I.1 Preliminary Remarks

In order to facilitate the practical application (both hand calculation and automatic computation) of the extensively discussed Computational Methods I, II, and III, this appendix contains a summary of the most important formulas, computational schemes, nomograms, and graphical and analytical representations of the universal functions and the conditions for starting the computations and their termination.

I.2 Computational Method I

(1) *Application*

This method is for laminar incompressible plane or axisymmetric boundary layers without heat transfer

(2) *Quantities given initially*

$u_\delta(x)$ free-stream velocity

$\Re(x)$ cross-sectional radius of the body of revolution

ν_∞ viscosity in the undisturbed flow.

$u_\delta(x)$ is approximated by a piecewise linear function. The velocity distribution in the interval $\Delta x = x_i - x_{i-1}$ is given by

$$u_\delta = u_{\delta_{i-1}} + \frac{u_{\delta_i} - u_{\delta_{i-1}}}{\Delta x} (x - x_{i-1}).$$

(3) *The unknowns in the boundary layer computation*

$Z = \delta_2 R_{\delta_2}$ thickness parameter

Γ shape parameter

(4) *The equations determining Z_i and Γ_i*

Momentum law:

$$\frac{Z_i}{Z_{i-1}}\left[\frac{\Re_i}{\Re_{i-1}}\right]^2 = A + B\frac{\Delta x}{Z_{i-1}}\cdot\frac{1+\left[\dfrac{\Re_i}{\Re_{i-1}}\right]^2}{2}.$$

Compatibility condition:

$$\left[\frac{\partial^2(u/u_\delta)}{\partial(y/\delta_2)^2}\right]_{y=0} = -\Gamma_i = -\frac{Z_i}{u_\delta}\frac{du_\delta}{dx}.$$

With the following definitions:

$$A = \left[\frac{u_{\delta_{i-1}}}{u_{\delta_i}}\right]^{b_l};$$

$$B = \frac{a_l}{1+b_l}\cdot\frac{1-\left[\dfrac{u_{\delta_{i-1}}}{u_{\delta_i}}\right]^{1+b_l}}{1-\dfrac{u_{\delta_{i-1}}}{u_{\delta_i}}},$$

where

$a_l = 0.441,$
$b_l = 4.165$ for $\Gamma > 0$ ⎫
$b_l = 5.165$ for $\Gamma < 0$ ⎬ Hartree profiles.

(5) *The solution of the equation for Z_i may also be obtained with the aid of the nomogram in Fig. 4.2, page 134.*

(6) *Various boundary layer quantities*

$$\delta_2 = \sqrt{Z\frac{\nu_\infty}{u_\delta}};$$

$$R_{\delta_2} = \frac{u_\delta\cdot\delta_2}{\nu_\infty};$$

$$\nu_\infty = \frac{\mu_\infty}{\rho_\infty}.$$

From Figs. I.1 and I.2 it can be seen that $\Gamma \to \beta^* \to H$ and thus

$$\delta_1 = \delta_2[4.0306 - 4.2845(H-1.515)^{0.3886}],$$

$$\delta = \frac{\delta_1}{\delta_1/\delta} \quad\text{where}\quad \frac{\delta_1}{\delta}(\Gamma) \quad\text{from Fig. I.3,}$$

$$c_f = (2/R_{\delta_2})\cdot 1.7261(H-1.515)^{0.7158}.$$

(7) Starting of the boundary layer computation

(a) plane flow

The leading edge of the body is approximated by a wedge with the angle $\beta^*\pi$. The shape parameter Γ which belongs to the angle $\beta^*\pi$ is obtained from Fig. I.2. The thickness parameter Z_i in the first integration interval $\Delta x = x_i$ is obtained from

$$Z_i = \frac{a_l}{1 + mb_l}x_i, \quad \text{where} \quad m = \frac{\beta^*}{2 - \beta^*}.$$

(b) axisymmetric flow

The leading edge of the axisymmetric body is approximated by a cone with the angle $\beta^*\pi$. As mentioned before, the flow about a cone can be reduced to plane flow by Mangler's coordinate transformation [70]. The angle of the wedge which is to be used for the corresponding plane flow is obtained from the cone angle by the following relations:

$$\beta_E^* = \frac{\beta^*}{3 - \beta^*}; \quad m_E = \frac{\beta_E^*}{2 - \beta_E^*}.$$

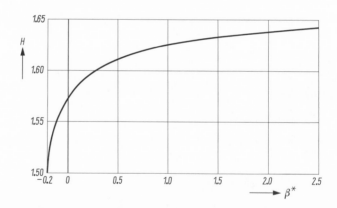

Fig. I.1. Function $H = H(\beta^*)$ for Hartree profiles with $\beta^* = 2m/(m+1) = -0.1988 + 2.30(H - 1.515) + 3800 [\exp (H - 1.515) - 1]^{3.85}$

From Fig. I.2, the shape parameter Γ of the boundary layer at the cone is obtained for any value of β_E^*. The thickness parameter Z of the conical flow in the first integration interval $\Delta x = x_i$ is obtained from

$$Z_i = \frac{a_l}{1 + m_E b_l}\frac{x_i}{3}, \quad \text{where} \quad m_E = \frac{\beta_E^*}{2 - \beta_E^*}; \quad \beta_E^* = \frac{\beta^*}{3 - \beta^*}.$$

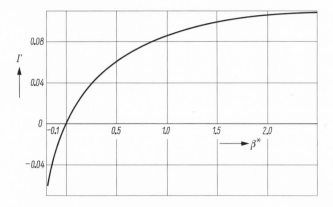

Fig. I.2.   Function $\Gamma = \Gamma(\beta^*) = \Gamma(m)$ for Hartree profiles; $m = \beta^*/(2 - \beta^*)$.

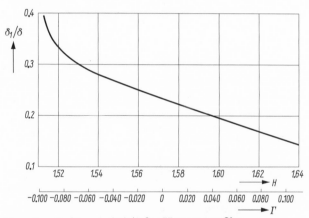

Fig. I.3.   Function $\delta_1/\delta = \delta_1(H)/\delta$ for Hartree profiles.

(8) Termination of the boundary layer computation

The laminar boundary layer computation is terminated either after

(a) the *laminar separation point* with $\Gamma_{\mathrm{sep}} = -0.0681$, or after

(b) the *laminar-turbulent transition,* a location that is approximated by the indifference point. The position of the indifference point for given parameters Γ, R_{δ_2} is obtained from Fig. I.4. If the point Γ, R_{δ_2} lies to the left of the critical curve in Fig. I.4, the subsequent computations have to be done with Computational Method II for turbulent boundary layers.

(9) Estimates for turbulent boundary layers

The reader is referred to the end of Section 4.2, page 135.

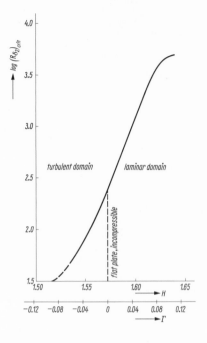

Fig. I.4. Function

$$(R_{\delta_2})_{\mathrm{crit}}(H)$$

for the determination of the indifference point. See also approximation, eq. (5.2).

I.3 Computational Method II

(1) Application

This method can be applied to laminar and turbulent, incompressible and compressible, plane and axisymmetric boundary layers, with and without heat transfer.

(2) The quantities given initially

$u_\delta(x)$ free-stream velocity or
$M_\delta(x)$ Mach number in the external flow,
$\Re(x)$ cross-sectional radius of the body of revolution,

$$\Theta(x) = \frac{T_e - T_w}{T_e - T_\delta} \text{ heat transfer parameter,}$$

M_∞ Mach number,

$R_\infty = \dfrac{\rho_\infty u_\infty L}{\mu_\infty}$ Reynolds number,

T_∞ absolute temperature in °K,

ρ_∞ density,

μ_∞ molecular viscosity,

in the approaching flow

The function $u_\delta(x)$ [or $M_\delta(x)$, respectively] is approximated by a piecewise linear function. The velocity distribution within the interval $\Delta x = x_i - x_{i-1}$ is here also given by

$$u_\delta = u_{\delta_{i-1}} + \frac{u_{\delta_i} - u_{\delta_{i-1}}}{\Delta x}(x - x_{i-1}),$$

or

$$M_\delta = M_{\delta_{i-1}} + \frac{M_{\delta_i} - M_{\delta_{i-1}}}{\Delta x}(x - x_{i-1}),$$

respectively.

(3) *The unknowns in the boundary layer computation*

$$Z = \delta_2 R_{\delta_2}^n = \delta_2 \left(\frac{\rho_\delta u_\delta \, \delta_2}{\mu_w}\right)^n = \text{thickness parameter}$$

$$H = \frac{(\delta_3)_u}{(\delta_2)_u} \qquad\qquad = \text{shape parameter}$$

(4) *Equations determining Z_i and H_i*

Momentum law

$$\frac{Z_i}{Z_{i-1}} \cdot \left(\frac{\mathfrak{R}_i}{\mathfrak{R}_{i-1}}\right)^{1+n} = A_Z + B_Z \overline{F}_2 \frac{\Delta x}{Z_{i-1}} \frac{1 + \left(\dfrac{\mathfrak{R}_i}{\mathfrak{R}_{i-1}}\right)^{1+n}}{2}$$

($n = 1$ laminar, $n = 0.268$ turbulent);

Energy law

$$\frac{H_i^*}{H_{i-1}^*} - 1 = \frac{\Delta H^*}{H_{i-1}^*} = A_H - 1 + B_H \overline{F}_4 \frac{2}{Z_i + Z_{i-1}} \cdot \frac{\Delta x}{H_{i-1}^*}$$

The shape parameter H is computed from the function $H = H^*(H, M_\delta, \Theta)$, given in 5(a) below. In the case where $M_\delta = 0$, $\Theta = 0$, we have $H^* = H$.

$$A_Z = \left(\frac{u_{\delta_{i-1}}}{u_{\delta_i}}\right)^{\overline{F}_1}; \qquad B_Z = \frac{1 - A_Z \dfrac{u_{\delta_{i-1}}}{u_{\delta_i}}}{(1 + \overline{F}_1)\left(1 - \dfrac{u_{\delta_{i-1}}}{u_{\delta_i}}\right)},$$

$$A_H = \left(\frac{u_{\delta_{i-1}}}{u_{\delta_i}}\right)^{\overline{F}_3}; \qquad B_H = \frac{1 - A_H \dfrac{u_{\delta_{i-1}}}{u_{\delta_i}}}{(1 + \overline{F}_3)\left(1 - \dfrac{u_{\delta_{i-1}}}{u_{\delta_i}}\right)}.$$

(5) *Universal functions for the method*

(a) relations that hold for *laminar* and *turbulent* boundary layers:

$$F_1(H, M_\delta, \Theta) = 2 + n + (1 + n)\frac{\delta_1}{\delta_2} - M_\delta^2,^\dagger$$

$$F_2(H, M_\delta, \Theta) = (1 + n)\frac{\delta_2}{(\delta_2)_u} \cdot \alpha(H),$$

$$F_3(H, M_\delta, \Theta) = 1 - \frac{\delta_1}{\delta_2} + 2\frac{\delta_4}{\delta_3},$$

$$F_4(H, M_\delta, \Theta) = \frac{\delta_2}{(\delta_2)_u}[2\beta R_{\delta_2}^{n-N} - \alpha H^*],$$

$$\frac{\delta_1}{\delta_2} = \frac{H_{12}}{\delta_2/(\delta_2)_u} + r\frac{\kappa - 1}{2}M_\delta^2(H^* - \Theta),$$

$$\frac{\delta_4}{\delta_3} = r\frac{\kappa - 1}{2}M_\delta^2\frac{H^* - \Theta}{H^*},$$

$$\frac{\delta_2}{(\delta_2)_u}(H, M_\delta, \Theta) = \frac{1}{1 + r\dfrac{\kappa - 1}{2}M_\delta^2(H^* - \Theta)(2 - H^*)},$$

$$\beta(H, M_\delta, \Theta) = \beta_u(H) \cdot \chi(M_\delta, \Theta),$$

$$H^*(H, M_\delta, \Theta) = H \cdot \psi(H, M_\delta, \Theta),$$

$$\psi(H, M_\delta, \Theta) = 1 + \frac{(\psi_{12} - 1)M_\delta}{M_\delta + \dfrac{\psi_{12} - 1}{\psi_0'}},$$

$$\psi_{12}(H, \Theta) = \frac{2 - (\delta_1)_u/\delta}{H} \cdot \Theta + \frac{1 - (\delta_1)_u/\delta}{H \cdot g}(1 - \Theta) = a - b\,\Theta,$$

$$\psi_0'(H, \Theta) = 0.0144(2 - H)(2 - \Theta)^{0.8}.$$

and from this:

$$H = \frac{1}{2\psi_{12}}\left[H^* + 2 + (\psi_{12} - 1)\left(2 + \frac{1}{0.0144\,M_\delta(2 - \Theta)^{0.8}}\right)\right]$$
$$+ \left\{\left|\frac{1}{4\psi_{12}^2}\left[H^* + 2 + (\psi_{12} - 1)\left(2 + \frac{1}{0.0144\,M_\delta(2 - \Theta)^{0.8}}\right)\right]^2\right.\right.$$
$$\left.\left. - \frac{H^*}{\psi_{12}}\left[2 + (\psi_{12} - 1)\frac{1}{0.0144\,M_\delta(2 - \Theta)^{0.8}}\right]\right\}^{1/2}.$$

$\dagger$If $\mu_w = \mu_w(x)$, i.e., $\mu_w' \neq 0$, then F_1 is replaced by
$$F_1^* = 2 + n + (1 + n)\,\delta_1/\delta_2 - M_\delta^2 + n\frac{\mu_w'/\mu_w}{u_\delta'/u_\delta} = F_1 + n\frac{\mu_w'/\mu_w}{u_\delta'/u_\delta}.$$

(b) relations that hold only for *laminar* boundary layers with $n = N = 1$ and $r = 0.85$, for air:

$$\alpha(H) = 1.7261(H - 1.515)^{0.7158},$$

$$\beta_u(H) = 0.1564 + 2.1921(H - 1.515)^{1.70},$$

$$H_{12}(H) = 4.0306 - 4.2845(H - 1.515)^{0.3886},$$

$$\left.\begin{array}{l} \dfrac{(\delta_1)_u}{\delta}(H) = 0.420 - (H - 1.515)^{0.424 \cdot H} \\[2mm] g(H) = 0.324 + 0.336(H - 1.515)^{0.555} \end{array}\right\} \quad \begin{array}{l} \text{where} \\[2mm] \delta = y\left(\dfrac{u}{u_\delta} = 0.999\right), \end{array}$$

$$a \approx 1.18$$
$$b \approx 1.08$$

$$\chi(M_\delta, \Theta, \omega) = \left[1 + \tfrac{2}{3}r\frac{\kappa - 1}{2}M_\delta^2(1 - \tfrac{3}{4}\Theta)\right]^\omega$$
$$\times \left\{1 + r\frac{\kappa - 1}{2}M_\delta^2(1 - \Theta)\right\}^{-\omega}, {}^*$$

or, a little more precisely,

$$\chi(M_\delta, \Theta, \omega, H)$$
$$= \left\{1 + r\frac{\kappa - 1}{2}M_\delta^2[(1.160H - 1.072) - \Theta(2H - 2.581)]\right\}^\omega$$
$$\times \left\{1 + r\frac{\kappa - 1}{2}M_\delta^2(1 - \Theta)\right\}^{-\omega}. {}^*$$

(c) relations that hold only for *turbulent* boundary layers with $n = 0.268$ and $r = 0.88$ for air:

$$(H) = 0.03894(H - 1.515)^{0.7},$$

$$(H_{\mathrm{sep}} = 1.515) = 0,$$

$$(H_{\mathrm{sep}})_{\mathrm{turb}} = 1.515,$$

$$\left.\begin{array}{l} \beta_u = 0.0056 \\ N = 0.168 \end{array}\right\} \text{ according to Rotta and Truckenbrodt,}\dagger$$

${}^*\omega = 0.5$ to about 1, depending on the range of temperatures. For wide ranges of the temperature one may put $\omega \approx 0.7$.
$\dagger$According to Felsch [27], the dissipation integral c_{Di} has to be computed from eq. (3.156), Section 3.7.3..

or

$$\beta_u(H) = 0.00481 + 0.0822(H - 1.5)^{4.81} \left.\vphantom{\begin{array}{c}a\\a\\a\end{array}}\right\} \begin{array}{l}\text{according}\\ \text{to Clauser}\\ \text{and Walz\dagger}\end{array}$$

$$N(H) = 0.2317H - 0.2644 - 0.87 \cdot 10^5(2 - H)^{20}$$

$$H_{12}(H) = 1 + 1.48(2 - H) + 104(2 - H)^{6.7},$$

$$\frac{(\delta_1)_u}{\delta}(H) = \frac{1}{2}\frac{2 - H}{H - 1} \left.\vphantom{\begin{array}{c}a\\a\end{array}}\right\} \begin{array}{l}\text{based on}\\ \text{eq. (3.123).}\end{array}$$

$$g(H) = 0.306 + (H - 1.5) - 0.885(H - 1.5)^{1.53},$$

$$\alpha(H_{\text{sep}}) = 0; \quad (H_{\text{sep}})_{\text{turb}} = 1.515$$

$$\alpha(H) = 0.03894\,(H - 1.515)^{0.7}$$

$$a \approx 1.09; \quad b \approx 0.05; \quad \chi = 1.0.$$

(6) Solution of the equations that determine Z_i and H_i

Z_i and H_i or H_i^*, respectively, are obtained by iterative solution of the two equations given under (4). Good convergence of this iterative method of solution is guaranteed so long as the following two conditions are satisfied on the integration interval Δx:

(a) $0.97 \le \dfrac{u_{\delta_i}}{u_{\delta_{i-1}}} \le 1.03,$

(b) $|\Delta H^*| = |H_i^* - H_{i-1}^*| \le 0.005.$

As a result of these conditions, the universal functions F_1 through F_4 may be replaced by constant averages $\overline{F}_1$ through $\overline{F}_4$ on the integration interval Δx. The average $\overline{F}$ is obtained as follows:

$$\overline{F} = F(\overline{H}; \overline{M}_\delta; \overline{\Theta})$$

$$= F\left(\frac{H_i + H_{i-1}}{2}; \ \frac{M_{\delta_i} + M_{\delta_{i-1}}}{2}; \ \frac{\Theta_i + \Theta_{i-1}}{2}\right).$$

Here $\overline{M}_\delta$, $\overline{\Theta}$ are immediately known from the problem. $\overline{H}$, on the other hand, has to be improved iteratively until the determining equations are satisfied. For the zeroth iteration, $H_i^{(0)}$ (in $\overline{H}$) is initially either estimated or one assumes $H_i^{(0)} = H_{i-1}$ (for automatic computation).

In the incompressible case (more generally, for constant material properties), this iterative method of solution can be carried out very quickly and simply with the aid of the nomograms in Figs. 4.3 through 4.7. For a boundary layer computation in the compressible case one uses

†According to Felsch [27], the dissipation integral c_{Di} has to be computed from eq. (3.156), Section 3.7.3.

the computational scheme A where it is advantageous to plot the universal functions graphically. For extensive boundary layer computations it is better to program the present method for an electronic computer.

(7) *Various boundary layer quantities*

$$\delta_2 = \sqrt[1+n]{Z\left(\frac{L}{R_\infty}\frac{u_\infty}{u_\delta}\frac{\rho_\infty}{\rho_\delta}\frac{\mu_w}{\mu_\infty}\right)^n},$$

$$R_{\delta_2} = \sqrt[1+n]{\frac{Z}{L}\cdot R_\infty}\cdot\frac{u_\delta}{u_\infty}\frac{\rho_\delta}{\rho_\infty}\frac{\mu_\infty}{\mu_w},$$

where

$$\frac{\rho_\infty}{\rho_\delta} = \left(\frac{T_\infty}{T_\delta}\right)^{\frac{1}{\kappa-1}}; \quad \frac{\mu_w}{\mu_\infty} = \left(\frac{T_w}{T_\infty}\right)^\omega,$$

$$\frac{T_\delta}{T_\infty} = \frac{1+\dfrac{\kappa-1}{2}M_\infty^2}{1+\dfrac{\kappa-1}{2}M_\delta^2},$$

$$\frac{T_w}{T_\infty} = \frac{1+r\dfrac{\kappa-1}{2}M_\delta^2(1-\Theta)}{1+\dfrac{\kappa-1}{2}M_\delta^2}\left(1+\frac{\kappa-1}{2}M_\infty^2\right),$$

$$\delta_1 = \delta_2\left[\frac{H_{12}}{\delta_2/(\delta_2)_u} + r\frac{\kappa-1}{2}M_\delta^2(H^*-\Theta)\right],$$

$$c_f = 2\cdot\frac{\alpha(H)}{R_{\delta_2}^n}\cdot\frac{\delta_2}{(\delta_2)_u}(H,M_\delta,\Theta),$$

see also 5(a), (b), (c).

(8) *Starting of the boundary layer computation*

(a) *plane laminar flow*

The same observations hold as in Computational Method I. The shape parameter H, which corresponds to the wedge angle $\beta^*\pi$, is obtained from Fig. I.1. So long as $\beta^*\pi \neq 0$, the shape parameter H for $x = 0$ can, also in the compressible case, be obtained from Fig. I.1, since in the immediate vicinity of the vertex of the wedge the flow may be considered incompressible. For $\beta^*\pi = 0$, that is, for flow along a flat plate, the shape parameter H in the compressible case has to be obtained from Fig. 6.5 for a given external velocity M_∞. The thickness parameter Z at $x = 0$ is equal to 0 in every case. In the first integration interval $\Delta x = x_i$, where wedge flow approximately holds, the following values for the initial parameters are found:

$$H_i = H_{i-1} = H = \text{const., from Fig. I.1}, \beta^* \neq 0,$$
$$H_i = H_{i-1} = H = \text{const., from Fig. 6.5}, \beta^* = 0,$$
$$Z_i = \frac{F_2}{1 + mF_1} \cdot x_i, \quad \text{where} \quad m = \frac{\beta^*}{2 - \beta^*}.$$

(b) axisymmetric laminar boundary layer

For incompressible and compressible flows with $0 < M_\infty < 1$, the same ideas prevail as in Method I. The shape parameter H which corresponds to the transformed angle $\beta_E^* \pi$ is again obtained from Fig. I.1.

In supersonic flow ($M_\infty > 1$), where only body shapes with a sharp bow are of practical interest, other considerations enter. As was shown in Section 6.2.6, the same flow patterns prevail along a flat plate and at a circular cone which approximates the pointed bow of the body to be investigated. The shape parameter H for $M_\infty > 1$ is therefore obtained from Fig. 6.5. This leads to the following initial values for the parameters in the first integration interval $\Delta x = x_i$:

$$H_i = H_{i-1} = H = \text{const., from Fig. I.1, for } M_\infty < 1$$
$$H_i = H_{i-1} = H = \text{const., from Fig. 6.5, for } M_\infty \geq 1$$
$$Z_i = \frac{F_2}{1 + m_E F_1} \frac{x_i}{3}, \quad \text{where} \quad m_E = \frac{\beta_E^*}{2 - \beta_E^*}; \quad \beta_E^* = \frac{\beta^*}{3 - \beta^*}.$$

(c) turbulent, plane or axisymmetric boundary layer

In this case, the start of the computations is always at the laminar turbulent transition point, whose position is obtained from I.2, 8(b). The initial values of the parameters H_{tur} and Z_{tur} of the turbulent computation at the transition point x_u are given as follows:

$$H_{\text{tur}} = H_{\text{lam}},$$
$$\delta_{2\,\text{tur}} = \delta_{2\,\text{lam}} \rightarrow Z_{\text{tur}} = \delta_{2\,\text{tur}} \cdot R_{\delta_2\,\text{tur}}^n.$$

(9) Termination of the boundary layer computation

(a) The *laminar* boundary layer computation is terminated either after the laminar separation point, characterized by $H_{\text{sep}} = 1.515$, or after *laminar-turbulent transition* whose position is determined as in Method I.

The parameters necessary to determine the indifference point from Fig. I.4 are here H and R_{δ_2}, however. The stability curve in Fig. I.4 may

Scheme A for Computational Method II, for laminar and turbulent boundary layers

⟨1⟩	⟨2⟩	⟨3⟩	⟨4⟩	⟨5⟩	⟨6⟩	⟨7⟩	⟨8⟩	⟨9⟩	⟨10⟩	⟨11⟩	⟨12⟩	⟨13⟩	⟨14⟩
x_i	u_{δ_i}	M_{δ_i}	Θ_i	$\dfrac{u_{\delta_i}}{u_\infty}\dfrac{\rho_{\delta_i}}{\rho_\infty}\dfrac{\mu_\infty}{\mu_{w_i}} R_\infty$	Δx	$\dfrac{u_{\delta_{i-1}}}{u_{\delta_i}}$	$\psi(M_{\delta_i}, \Theta_i)$	$\overline{M}_\delta$	$\overline{\Theta}$	$\overline{H}$	$\overline{F}_1$	$\overline{F}_3$	$A_z \cdot Z_{i-1}$
					$x_i - x_{i-1}$			$\overline{M}_\delta = \dfrac{M_{\delta_i} + M_{\delta_{i-1}}}{2}$	$\overline{\Theta} = \dfrac{\Theta_i + \Theta_{i-1}}{2}$	$\overline{H} = \dfrac{H_i + H_{i-1}}{2}$	$\overline{F}_1(\overline{M}_\delta, \overline{\Theta}, \overline{H})$	$\overline{F}_3(\overline{M}_\delta, \overline{\Theta}, \overline{H})$	

⟨1⟩–⟨6⟩: given values   ⟨9⟩–⟨14⟩: iterative improvement

⟨15⟩	⟨16⟩	⟨17⟩	⟨18⟩	⟨19⟩	⟨20⟩	⟨21⟩	⟨22⟩	⟨23⟩	⟨24⟩	⟨25⟩	⟨26⟩	⟨27⟩	⟨28⟩
$A_H \cdot H^*_{i-1}$	B_z	B_H	$\overline{F}_2$	$B_z \overline{F}_2 \Delta x$	Z_i	$Z = \dfrac{Z_i + Z_{i-1}}{2}$	$\overline{F}_4$	$B_H \overline{F}_4 \dfrac{\Delta x}{Z}$	H^*	H_i	R_{δ_2}	c_{f_i}	δ_{2_i}
			$\overline{F}_2(\overline{M}_\delta, \overline{\Theta}, \overline{H})$	$\langle16\rangle \cdot \langle18\rangle \cdot \langle6\rangle$	$\langle14\rangle + \langle19\rangle$		laminar: $\overline{F}_4(\overline{M}_\delta, \overline{\Theta}, \overline{H})$ turbulent: $\overline{F}_4(\overline{M}_\delta, \overline{\Theta}, \overline{R}_{\delta_2}, \overline{H})$	$\dfrac{\langle17\rangle \cdot \langle22\rangle \cdot \langle6\rangle}{\langle21\rangle}$	$\langle15\rangle + \langle23\rangle$	$H_i = H_i(H^*_i, \psi)$	$(\langle20\rangle \cdot \langle5\rangle)^{\frac{1}{1+n}}$	$\dfrac{F_{2_i}}{(1 + n)\langle26\rangle^n}$	$\dfrac{\langle20\rangle}{\langle26\rangle^n}$

⟨18⟩ ff.: given values   iterative improvement

For $M = 0$, $H^* = H$ holds

Explanations: The initial value $(\ )_{i-1}$ is the end value $(\ )_i$ of the previous step. For the zeroth iteration the shape parameter then be approximately applied to compressible boundary layer as well.

(b) The computation of the *turbulent* boundary layer is terminated after turbulent separation which is characterized by $H_{\text{sep}} = 1.50$ to 1.57.

I.4 Computational Method III

(1) Application

This method is to be applied to *laminar*, incompressible and compressible, plane and axisymmetric boundary layers, with and without heat transfer.

(2) The variables of the problem

$u_\delta(x)$ Free-stream velocity, or

$M_\delta(x)$ Mach number in the free-stream flow, or

$M_\delta^*(x) = f(M_\delta) = $ critical Mach number in the free-stream flow (see Appendix II)

$\Re(x)$ Radius of cross section of body of revolution

$$\left.\begin{array}{l} \Theta(x) = \dfrac{T_e - T_w}{T_e - T_\delta} \text{ when } M_\infty \neq 0 \\[1.5em] c_T(x) = \dfrac{T_\delta - T_w}{T_\delta} \text{ when } M_\infty = 0\dagger \end{array}\right\} \text{ Heat-transfer parameters}$$

$$\left.\begin{array}{l} M_\infty \text{ Mach number} \\[0.8em] R_\infty = \dfrac{\rho_\infty u_\infty L}{\mu_\infty} \text{ Reynolds number} \\[0.8em] T_\infty \text{ absolute temperature in } {}^\circ K \\[0.5em] \rho_\infty \text{ density} \\[0.5em] \mu_\infty \text{ dynamic viscosity} \end{array}\right\} \text{ in the approaching flow}$$

(3) The unknowns of the boundary layer computation

$$\left.\begin{array}{l} Z_J = \dfrac{u_\delta}{\nu_\infty} \cdot \Delta_2^2 = \Delta_2 \cdot R_{\Delta_2} \\[1.2em] Z_E = \dfrac{u_\delta}{\nu_\infty} \cdot \Delta_2 \cdot \Delta_3 = \Delta_3 \cdot R_{\Delta_2} \end{array}\right\} \text{ Thickness parameters}$$

$$\left.\begin{array}{l} H^* = \dfrac{Z_E}{Z_J} \\[1em] \Gamma^* \end{array}\right\} \text{ Shape parameters}$$

$\dagger \; \displaystyle\lim_{M_\delta \to 0}\left[\Theta \cdot \frac{\kappa - 1}{2} M_\delta^2\right] = c_T.$

(4) The equations for Z_{J_i}, Z_{E_i}, and Γ_i^*.

Momentum law

$$\left. \begin{array}{l} \dfrac{(Z_J)_i}{(Z_J)_{i-1}} \cdot \dfrac{\mathfrak{R}_i^2}{\mathfrak{R}_{i-1}^2} = A_J + \bar{F}_J^* \dfrac{B_J}{(Z_J)_{i-1} \cdot \mathfrak{R}_{i-1}^2} \\[3mm] \text{Energy law} \\[2mm] \dfrac{(Z_E)_i}{(Z_E)_{i-1}} \cdot \dfrac{\mathfrak{R}_i^2}{\mathfrak{R}_{i-1}^2} = A_E + \bar{F}_E^* \dfrac{B_E}{(Z_E)_{i-1} \cdot \mathfrak{R}_{i-1}^2} \end{array} \right\} \quad H_i^* = \dfrac{(Z_E)_i}{(Z_J)_i},$$

Compatibility condition

$$\left[\frac{\partial^2 u/u_\delta}{\partial (Y/\Delta_2)^2} \right]_{Y=0} = -\Gamma_i^* = -(Z_J)_i \frac{1}{u_{\delta_i}} \frac{du_{\delta_i}}{dx} \frac{T_w}{T_\delta}$$

$$= -(Z_J)_i \frac{1}{\mathrm{M}_{\delta_i}^*} \frac{d\mathrm{M}_{\delta_i}^*}{dx} \frac{T_w}{T_\delta}.$$

The following quantities depend only on the initial data: For $\mathrm{M}_\infty = 0$ without or with heat transfer

$$A_J = \left[\frac{u_{\delta_{i-1}}}{u_{\delta_i}} \right]^{4-\bar{c}_T}, \qquad B_J = \frac{\displaystyle\int_{x_{i-1}}^{x_i} u_\delta^{4-c_T} \cdot \mathfrak{R}^2 \, dx}{u_{\delta_i}^{4-c_T}},$$

$$A_E = \left[\frac{u_{\delta_{i-1}}}{u_{\delta_i}} \right]^{4-1.27\cdot\bar{c}_T}, \qquad B_E = \frac{\displaystyle\int_{x_{i-1}}^{x_i} u_\delta^{4-1.27\cdot c_T} \cdot \mathfrak{R} \, dx}{u_{\delta_i}^{4-1.27\cdot c_T}},$$

where

$$\bar{c}_T = \tfrac{1}{2}[c_{T_i} + c_{T_{i-1}}].$$

For $\mathrm{M}_\infty \neq 0$, without or with heat transfer

$$A = \left[\frac{\mathrm{M}_{\delta_{i-1}}^*}{\mathrm{M}_{\delta_i}^*} \right]^4 \left[\frac{1 - \dfrac{\kappa - 1}{\kappa + 1} \mathrm{M}_{\delta_{i-1}}^{*2}}{1 - \dfrac{\kappa - 1}{\kappa + 1} \mathrm{M}_{\delta_i}^{*2}} \right]^a,$$

$$B = \frac{\displaystyle\int_{x_{i-1}}^{x_i} \mathrm{M}_\delta^{*4} \left(1 - \frac{\kappa - 1}{\kappa + 1} \mathrm{M}_\delta^{*2} \right)^a \cdot \mathfrak{R}^2 \cdot dx}{\mathrm{M}_{\delta_i}^{*4} \left(1 - \dfrac{\kappa - 1}{\kappa + 1} \mathrm{M}_{\delta_i}^{*2} \right)^a}, *$$

*The integral in the quantity B can be obtained graphically very rapidly and with sufficient accuracy.

where

$$\left.\begin{array}{l} A_J = A \\ B_J = B \end{array}\right\} \quad \text{for} \quad a = -\frac{r}{2}(3 - \overline{\Theta}) + \frac{\kappa}{\kappa - 1},$$

$$\left.\begin{array}{l} A_E = A \\ B_E = B \end{array}\right\} \quad \text{for} \quad a = -\frac{r}{2}(3 - 1.27\overline{\Theta}) + \frac{\kappa}{\kappa - 1},$$

and

$$\overline{\Theta} = \tfrac{1}{2}[\Theta_i + \Theta_{i-1}],$$

$$\frac{T_w}{T_\delta} = 1 + r\frac{\kappa - 1}{2} M_\delta^2[1 - \Theta],$$

$$\kappa = 1.4 \quad \text{for air}$$

$$r = 0.85 \quad \text{laminar.}$$

(5) *The universal functions of the method*

$$F_J^*(H^*, \Gamma^*, M_\delta, \Theta) = F_J(H^*, \Gamma^*) + \left[\frac{\partial \frac{u}{u_\delta}}{\partial Y/\Delta_2}\right]_{Y=0} \cdot \left[2\frac{\mu_w}{\mu_\infty}\frac{T_\infty}{T_w} - 2\right],$$

$$F_E^*(H^*, \Gamma^*, M_\delta, \Theta) = F_E(H^*, \Gamma^*) \cdot \frac{\mu_m}{\mu_\infty}\frac{T_\infty}{T_m}$$
$$+ H^*[F_J(H^*, \Gamma^*) - \Gamma^*] \cdot \frac{1}{2}\left[\frac{\mu_w}{\mu_\infty}\frac{T_\infty}{T_w} - \frac{\mu_m}{\mu_\infty}\frac{T_\infty}{T_m}\right],$$

where

$F_J(H^*; \Gamma^*)$ from Fig. 4.8

$F_E(H^*; \Gamma^*)$ from Fig. 4.9

$$\left[\frac{\partial \frac{u}{u_\delta}}{\partial Y/\Delta_2}\right]_{Y=0} = f(H^*; \Gamma^*) \quad \text{from Fig. 4.10}$$

$$\frac{T_w}{T_\infty} = \frac{1 + r\frac{\kappa - 1}{2} M_\delta^2(1 - \Theta)}{1 + \frac{\kappa - 1}{2} M_\delta^2}\left(1 + \frac{\kappa - 1}{2} M_\infty^2\right),$$

$$\frac{T_m}{T_\infty} = \frac{1 + r\frac{\kappa - 1}{2} M_\delta^2(\frac{2}{3} - \frac{1}{2}\Theta)}{1 + \frac{\kappa - 1}{2} M_\delta^2}\left(1 + \frac{\kappa - 1}{2} M_\infty^2\right),$$

$$\mu = \mu(T) \begin{cases} \dfrac{\mu}{\mu_\infty} = \left(\dfrac{T}{T_\infty}\right)^\omega & \text{where} \quad 0.5 < \omega < 1, \\[4mm] \dfrac{\mu}{\mu_\infty} = \left(\dfrac{T}{T_\infty}\right)^{3/2} \dfrac{1 + \dfrac{S}{T_\infty}}{T/T_\infty + S/T_\infty} & \text{where} \quad S = 110°\text{K}. \end{cases}$$

Special case:

$$\left.\begin{array}{l} F_J^* = F_J(H^*, \Gamma^*) \\ F_E^* = F_E(H^*, \Gamma^*) \end{array}\right\} \quad \text{for} \quad \mu \sim T; \quad \omega = 1.$$

(6) *Solution of the equation for Z_{J_i}, Z_{E_i} and Γ_i^**

The quantities Z_{J_i}, Z_{E_i}, or Γ_i^* and H_i^* are obtained by solving the equations iteratively. Good convergence of this method is guaranteed if the same conditions as in Method II are satisfied inside the integration interval $\Delta x = x_i - x_{i-1}$,

$$0.97 \leq \frac{u_{\delta_i}}{u_{\delta_{i-1}}} \leq 1.03,$$

$$|\Delta H^*| = |H_i^* - H_{i-1}^*| \leq 0.005.$$

For the other variables one proceeds analogous to (6) in I.3. It should be noted that the two universal functions F_J^* and F_E^* of the present method depend on two shape parameters, H^* and Γ^*, and that therefore both averages $\overline{H}^*$ and $\overline{\Gamma}^*$ have to be improved iteratively. For purposes of rapidly carrying out the interative method it is useful to employ computational scheme B and the graphs of the universal functions in Figs. 4.8, 4.9, and 4.10.

(7) *Various boundary layer parameters*

$$\delta_2 = \Delta_2 \frac{T_\delta}{T_\infty} \cdot \left(\frac{p_\infty}{p}\right)^{1/2}$$

where

$$\Delta_2 = \sqrt{Z_J \frac{\nu_\infty}{u_\delta}},$$

$$\frac{T_\delta}{T_\infty}\left(\frac{p_\infty}{p}\right)^{1/2} = \left[\frac{1 + \dfrac{\kappa - 1}{2} M_\delta^2}{1 + \dfrac{\kappa - 1}{2} M_\infty^2}\right]^{\frac{2-\kappa}{\kappa - 1}}$$

$$R_{\delta_2} = \frac{\rho_\delta u_\delta \delta_2}{\mu_w} = \sqrt{\frac{u_\delta}{\nu_\infty}} Z_J \cdot C_R$$

$$C_R = \frac{\mu_\delta}{\mu_w} \cdot \frac{T_\delta}{T_\infty} \cdot \left(\frac{p_\infty}{p}\right)^{1/2},$$

$$c_f = 2 \frac{\left[\frac{\partial u/u_\delta}{\partial Y/\Delta_2}\right]_{Y=0} \cdot \frac{T_\delta}{T_w} \frac{\mu_w}{\mu_\delta}}{\sqrt{\frac{u_\delta}{\nu_\infty}} Z_J \cdot \frac{T_\delta}{T_\infty} \left(\frac{p_\infty}{p}\right)^{1/2}}$$

and

$$\left[\frac{\partial u/u_\delta}{\partial Y/\Delta_2}\right]_{Y=0} = f(H^*, \Gamma^*) \quad \text{from Fig. 4.10}$$

$$\frac{T_w}{T_\delta} \quad \text{and} \quad \frac{\mu_w}{\mu_\delta}, \quad \text{see (5).}$$

Additional relations can be found in reference [33].

(8) *Starting of the boundary layer computation*

(a) *plane laminar flow*

Analogous to the considerations under 8(a) in I.3, the initial values of the shape parameters H^* and Γ^* are transformed to those of the corresponding flow along a wedge with the angle $\beta^*\pi \neq 0$. They can be obtained from Figs. I.1 and I.2.

For $\beta^*\pi = 0$, that is, for a flow along a flat plate, the shape parameters are obtained as follows:

$$\left.\begin{array}{l} \Gamma^* = 0 \quad \text{since} \quad \dfrac{du_\delta}{dx} = 0 \\[2ex] H^* = \dfrac{F_E^*(H^*; \Gamma^* = 0; M_\delta; \Theta)}{F_J^*(H^*; \Gamma^* = 0; M_\delta; \Theta)} \end{array}\right\} \begin{array}{l} \text{for arbitrary } \mu(T)\text{-law,} \\ H^* \text{ is found by iteration.} \end{array}$$

$$\left.\begin{array}{l} \Gamma^* = 0 \\ H^* = 1.572 \end{array}\right\} \text{for } \mu \sim T.$$

The thickness parameters in the first integration interval $\Delta x = x_i$ are:

$$\left.\begin{array}{l} (Z_J)_i = \overline{F}_J^* \cdot B_J \\ (Z_E)_i = \overline{F}_E^* \cdot B_E \\ H_i^* = \text{const.}; \quad \Gamma_i^* = \text{const.} \end{array}\right\} \text{for } \beta^* \neq 0,$$

$$\left.\begin{array}{l} (Z_J)_i = \overline{F}_J^* x_i \\[4pt] (Z_E)_i = \overline{F}_E^* x_i \\[4pt] H_i^* = \text{const.}; \quad \Gamma_i^* = 0 \end{array}\right\} \quad \text{for} \quad \beta^* = 0.$$

(b) axisymmetric laminar flow

The same remarks hold as under 8(b) in I.3 for Method II. The initial values of the parameters in the first integration interval $\Delta x = x_i$ are then given by

$$\left.\begin{array}{l} H_i^* = \text{const.,} \quad \text{for } \beta_E^* \text{ from Fig. I.1} \\[6pt] \Gamma_i^* = \text{const.,} \quad \text{for } \beta_E^* \text{ from Fig. I.2} \\[6pt] (Z_J)_i = \overline{F}_J^* \dfrac{B_J}{\mathfrak{R}_i^2} \\[10pt] (Z_E)_i = \overline{F}_E^* \dfrac{B_E}{\mathfrak{R}_i^2} \end{array}\right\} \quad \text{for } M_\infty < 1,$$

$$\left.\begin{array}{l} H_i^* = \dfrac{F_E^*(H_i^*, \Gamma_i^* = 0, M_\delta, \Theta)}{F_J^*(H_i^*, \Gamma_i^* = 0, M_\delta, \Theta)} \\[12pt] \Gamma_i^* = 0 \\[6pt] (Z_J)_i = \overline{F_J^*} \cdot \dfrac{x_i}{3} \\[10pt] (Z_E)_i = \overline{F_E^*} \cdot \dfrac{x_i}{3} \end{array}\right\} \quad \text{for } M_\infty > 1.$$

(9) Termination of the boundary layer computation

The computation of laminar boundary layers is terminated either after the *laminar separation point*, which is characterized by

$$\left[\frac{\partial \dfrac{u}{u_\delta}}{\partial Y/\Delta_2}\right]_{Y=0} = f(H_{\text{sep}}^*, \Gamma_{\text{sep}}^*) = 0$$

(see Fig. 4.10) or after the *laminar-turbulent transition*, whose location is approximated by the indifference point. The position of the indifference point for the computed parameters H^*, R_{Δ_2} can be obtained from Fig. I.4, if one puts approximately, also for compressible flow, $H^* = H$ and $R_{\Delta_2} = R_{\delta_2}$. Other than that, the same remarks as in Methods I and II hold for the determination of the indifference point.

Scheme B for Computational Method III, for laminar boundary layers

	⟨1⟩	⟨2⟩	⟨3⟩	⟨4⟩	⟨5⟩	⟨6⟩	⟨7⟩	⟨8⟩	⟨9⟩	⟨10⟩	⟨11⟩	⟨12⟩	⟨13⟩
symbol	x_i	$M_{\delta_i}^*$	Θ_i	$\dfrac{T_w}{T_\delta}$	$\dfrac{1}{M_\delta^*}\dfrac{dM_\delta^*}{dx}\dfrac{T_w}{T_\delta}$	Δx	A_J	B_J	A_E	B_E	$(Z_J)_{i-1}\cdot A_J$	$(Z_E)_{i-1}\cdot A_E$	$\overline{H}^* = \dfrac{H_{i-1}^* + H_i^*}{2}$
formula					*given values*	$x_i - x_{i-1}$	For $M_\infty = 0$, $\Theta = 0$: ⟨7⟩ = ⟨9⟩ and ⟨8⟩ = ⟨10⟩				$\langle 19\rangle_{i-1}\cdot\langle 7\rangle$	$\langle 20\rangle_{i-1}\cdot\langle 9\rangle$	$\dfrac{\langle 21\rangle_{i-1} + \langle 21\rangle_i}{2}$

iterative improvement

	⟨14⟩	⟨15⟩	⟨16⟩	⟨17⟩	⟨18⟩	⟨19⟩	⟨20⟩	⟨21⟩	⟨22⟩	⟨23⟩	⟨24⟩	⟨25⟩	⟨26⟩	⟨27⟩	⟨28⟩
symbol	$\overline{\Gamma}^* = \dfrac{\Gamma_{i-1}^* + \Gamma_i^*}{2}$	$\overline{F}_J^*$	$\overline{F}_E^*$	$\overline{F}_J^*\cdot B_J$	$\overline{F}_E^*\cdot B_E$	$(Z_J)_i$	$(Z_E)_i$	H_i^*	Γ_i^*	$\left[\dfrac{\partial u/u_\delta}{\partial Y/\Delta_2}\right]_w$	$\left[\dfrac{\partial u/u_\delta}{\partial y/\delta_2}\right]_w$	C_R	R_{δ_2}	c_f	δ_2
formula	$\dfrac{\langle 22\rangle_{i-1} + \langle 22\rangle_i}{2}$	$\overline{F}_J(\overline{H}^*, \Gamma^*)$	$\overline{F}_E^*(\overline{H}^*, \overline{\Gamma}^*)$	$\langle 15\rangle\cdot\langle 8\rangle$	$\langle 16\rangle\cdot\langle 10\rangle$	$\langle 11\rangle + \langle 17\rangle$	$\langle 12\rangle + \langle 18\rangle$	$\dfrac{\langle 20\rangle}{\langle 19\rangle}$	$\langle 5\rangle\cdot\langle 19\rangle$	from Fig. 4.10	$\dfrac{\langle 23\rangle}{\langle 4\rangle}$	see Appendix I.4	$\sqrt{\langle 19\rangle}\cdot\langle 25\rangle$	$\dfrac{\langle 24\rangle}{\langle 26\rangle}$	see Appendix I.4

iterative improvement

Explanations: The initial value $(\ )_{i-1}$ is the end value $(\ )_i$ of the previous step. For the zeroth iteration the shape parameter is put $\overline{H}^* = H_{i-1}$ and $\overline{\Gamma}^* = \Gamma_{i-1}^*$. For the definition of the various quantities, see Section I.4.

I.5 Computation of the Heat Transfer

For Computational Methods II and III, the heat energy $q_w(x)$ exchanged between the wall and the boundary layer at a point x is obtained from

$$q_w = - \frac{\rho_\delta}{2} u_\delta^3 \frac{r}{s} \frac{c_f}{2} \cdot \Theta \quad \text{for } M_\delta > 0$$

or

$$q_w = - \frac{\rho_\delta}{2} u_\delta^3 \frac{c_f}{2s} b \frac{c_p T_\delta}{u_\delta^2/2} \quad \text{for } M_\delta = 0.$$

The total heat energy exchanged between $x = 0$ and $x = L$ is then

$$Q = \int_0^L q_w(x)\, \mathrm{d}x.$$

A positive sign of q_w and Q indicates a heat flux in positive y direction, that is, from the wall into the boundary layer.

For a more precise computation of the heat transfer in flows with strong temperature and pressure gradients in the direction of flow ($\mathrm{d}T_w/\mathrm{d}x \neq 0$, $\mathrm{d}p/\mathrm{d}x \neq 0$) we have

$$q_w = - \frac{\rho_\delta}{2} u_\delta^3 \frac{r}{s} \frac{c_f}{2} \cdot \Theta \left[1 + \frac{K}{b} \right] \quad \text{for } M_\delta > 0$$

or

$$q_w = - \frac{\rho_\delta}{2} u_\delta^3 \frac{c_f}{2s} [b + K] \frac{c_p T_\delta}{u_\delta^2/2} \quad \text{for } M_\delta = 0.$$

Here the function $K(x)$ has to be computed according to Sect. 5.3.

Appendix II

Important Gas Dynamic Relations

In the following some gas-dynamical relations often used in boundary layer theory are brought together. Detailed derivations of these relations can be found, for example, in the books by Oswatitsch [85], Sauer [110], and Zierep [160].

It is generally assumed for compressible boundary layers that the gases under consideration, such as air, behave approximately like ideal gases and that the specific heats are constant. These assumptions are valid for pressure and temperatures that are not too extreme.

(1) Equation of state for ideal gases

$$p = \rho RT.$$

(2) Material properties

Universal gas constant R in $\mathrm{m^2\,s^{-2}\,^\circ K^{-1}}$
Specific heat at constant pressure c_p in $\mathrm{m^2\,s^{-2}\,^\circ K^{-1}}$
Specific heat at constant volume c_v in $\mathrm{m^2\,s^{-2}\,^\circ K^{-1}}$
Adiabatic exponent $\kappa = c_\mathrm{p}/c_\mathrm{v}$ $(= 1.405$ for air)

(3) Velocities in $\mathrm{m/s}$

Local velocity of sound $a = \sqrt{\dfrac{\kappa p}{\rho}} = \sqrt{\kappa RT}$,

Velocity of sound at rest (subscript 0), that is, for $u = 0$: $T = T_0$; $p = p_0$; $\rho = \rho_0$.

$$a_0 = \sqrt{\frac{\kappa p_0}{\rho_0}} = \sqrt{\kappa RT_0},$$

Critical velocity of sound $a^* = a_0 \cdot \sqrt{\dfrac{2}{\kappa + 1}}$,

Local velocity u

(4) *Ratios of velocities*

Local Mach number M $= u/a$

Critical local Mach number M* $= u/a*$

(M$_{max}^*$ $= 2.437$ for $\kappa = 1.405$ corresponds to M$_{max} = \infty$)

$$M^2 = \frac{2}{\kappa + 1} \cdot \frac{M^{*2}}{1 - \dfrac{\kappa - 1}{\kappa + 1} M^{*2}},$$

$$M^{*2} = \frac{\kappa + 1}{2} \frac{M^2}{1 + \dfrac{\kappa - 1}{2} M^2}.$$

(5) *Relations between pressure, density, temperature and Mach number for adiabatic changes of state*

(Subscript 0 = quantities at rest; Subscript ∞ = quantities in the external flow; Subscript δ = quantities at the edge of the boundary layer; $p = p_\delta$)

$$\frac{p}{p_0} = \left[\frac{\rho_\delta}{\rho_0}\right]^\kappa,$$

$$\frac{p}{p_0} = \left[\frac{T_0}{T_\delta}\right]^{\frac{\kappa}{1-\kappa}},$$

$$\frac{p}{p_0} = \left[1 + \frac{\kappa - 1}{2} M_\delta^2\right]^{\frac{\kappa}{1-\kappa}} = \left[1 - \frac{\kappa - 1}{\kappa + 1} M_\delta^{*2}\right]^{\frac{\kappa}{\kappa-1}},$$

$$\frac{p_0}{p_\infty} = \left[1 + \frac{\kappa - 1}{2} M_\infty^2\right]^{\frac{\kappa}{\kappa-1}} = \left[1 - \frac{\kappa - 1}{\kappa + 1} M_\infty^{*2}\right]^{\frac{\kappa}{1-\kappa}},$$

$$\frac{\rho_0}{\rho_\delta} = \left[\frac{T_0}{T_\delta}\right]^{\frac{1}{\kappa-1}} \quad \text{with} \quad \rho_0 = \frac{\kappa p_0}{a_0^2},$$

$$\frac{1}{\rho_\delta} \frac{d\rho_\delta}{dx} = - \frac{1}{u_\delta} \frac{du_\delta}{dx} \cdot M_\delta^2,$$

$$\frac{T_0}{T_\delta} = 1 + \frac{\kappa - 1}{2} M_\delta^2 = \frac{1}{1 - \dfrac{\kappa - 1}{\kappa + 1} M_\delta^{*2}},$$

$$\frac{T_0}{T_\infty} = 1 + \frac{\kappa - 1}{2} M_\infty^2 = \frac{1}{1 - \dfrac{\kappa - 1}{\kappa + 1} M_\infty^{*2}},$$

$$M_\delta^2 = \frac{2}{\kappa - 1}\left[\left(\frac{p_0}{p}\right)^{\frac{\kappa-1}{\kappa}} - 1\right],$$

$$M_\delta^{*2} = \frac{\kappa + 1}{\kappa - 1}\left[1 - \left(\frac{p}{p_0}\right)^{\frac{\kappa-1}{\kappa}}\right].$$

(6) *Ratio of the Kinetic Energy $u_\delta^2/2$ per unit of mass to the enthalpy* $i_\delta = c_p T_\delta$

$$\frac{u_\delta^2/2}{i_\delta} = \frac{1}{2}M_\delta^2\frac{a_\delta^2}{c_p T_\delta} = \frac{M_\delta^2}{2}\frac{\kappa R T_\delta}{c_p T_\delta} = \frac{\kappa - 1}{2}M_\delta^2.$$

Appendix III
Flow Diagram for
Computational Method II

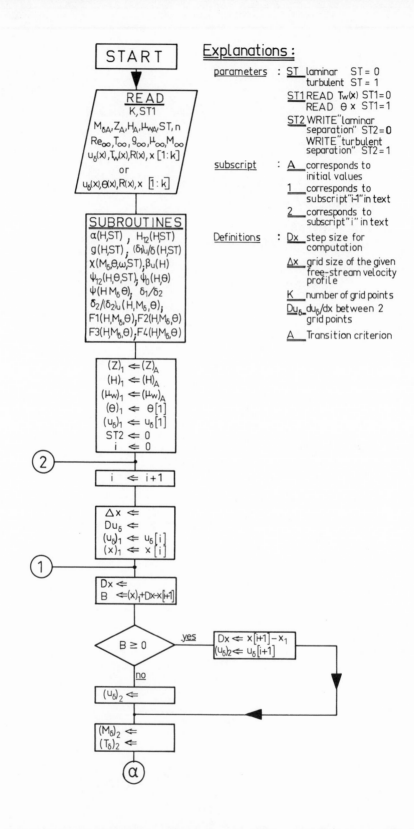

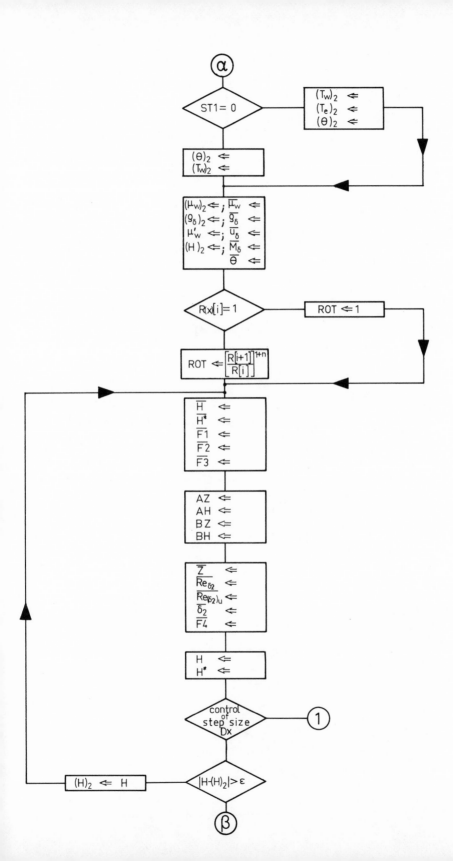

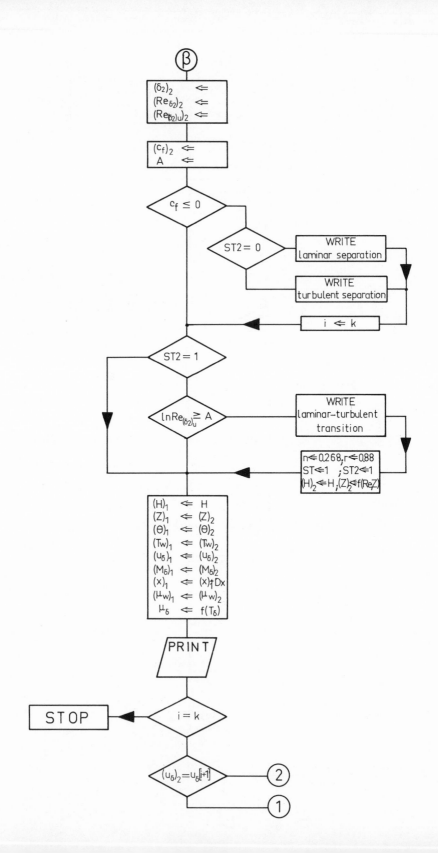

References

[1] J. Ackeret: Grenzschichten in geraden und gekrümmten Diffusoren. In: Grenz-schichtforschung, IUTAM-Symposium Freiburg i. Br. 1957, H. Görtler, ed., Springer-Verlag, Berlin, Göttingen, Heidelberg 1958.

[2] R. Betchov, W. O. Criminale: On the oscillations of a turbulent flow. Rep. No. ATN-64 (9231)-1 of the Aerospace Corp. El Segundo, California (1964).

[3] A. Betz: Konforme Abbildung. Springer-Verlag, Berlin, Göttingen, Heidelberg 1964.

[4] H. Blasius: Grenzschichten in Flüssigkeiten mit kleiner Reibung. Z. für Math. u. Phys. 56, 1 (1908).

[5] A. Buri: Eine Berechnungsgrundlage für die turbulente Grenzschicht bei beschleu-nigter und verzögerter Grundströmung. Thesis, Fed. Inst. of Tech. Zürich 1931.

[6] A. Busemann: Gasströmung mit laminarer Grenzschicht entlang einer Platte. Z. f. angew. Math u. Mech. 15, 23–25 (1935).

[7] D. R. Chapman, M. W. Rubesin: Temperature and velocity profiles in the com-pressible laminar boundary layer with arbitrary distribution of surface tempera-ture. J. Aeron. Sc. 16, 547–565 (1949).

[8] F. H. Clauser: Turbulent boundary layers in adverse pressure gradients. J. Aeron. Sc. 21, 91–108 (1954).

[9] D. Coles: The law of the wake in the turbulent boundary layer. J. Fluid Mech. 1, 191–226 (1956).

[10] L. Crocco: Sullo strato limite laminare nei gas lungo una lamina plana. Rend. Math. Univ. Roma V 2, 138 (1941).

[11] W. Dienemann: Berechnung des Wärmeübergangs an laminar umströmten Körpern mit konstanter und ortsveränderlicher Wandtemperatur. Thesis, Tech-nical University Braunschweig 1951; Z. f. Math. u. Mech. 33, 89–109 (1953).

[12] E. A. von Doenhoff, H. Tetervin: Determination of general relations for the behaviour of turbulent boundary layers. NACA Rep. 772 (1943).

[13] A. A. Dorodnitsyn: Boundary layer in a compressible gas. Prikladnaya Mathe-matika i Mechanika VI (1942).

[14] A. A. Dorodnitsyn: General method of integral relations and its application to boundary layer theory. Advances in Aeron. Sc., Vol. 3. Proceedings of the 2nd Intern. Congr. on the Aeron. Sc., Zürich 1960. Pergamon Press, London, 1962, pp. 207–219.

[15] E. R. van Driest: Calculation of the stability of the laminar boundary layer in a compressible fluid on a flat plate with heat transfer. J. Aeron. Sc. 19, 801–813 (1952).

[16] E. R. van Driest, J. C. Boison: Boundary layer stabilization by surface cooling in supersonic flow. J. Aeron. Sc. 22, 70 (1955).

[17] E. R. van Driest in: C. C. Lin: Turbulent flows and heat transfer, Sect. F. Princeton Univ. Press, 1959.

[18] D. W. Dunn, C. C. Lin: On the stability of the laminar boundary layer in a compressible fluid. J. Aeron. Sc. 22, 455–477 (1955); see also J. Aeron Sc. 20, 577 (1953) and 19, 491 (1952).

[19] E. R. G. Eckert: Einführung in den Wärme- und Stoffaustausch. Springer-Verlag, Berlin, 1966.

[20] E. R. G. Eckert, R. Eichhorn, Th. L. Eddy: Measurement of temperature profiles in laminar and turbulent axisymmetric boundary layers on a cylinder with nonuniform wall temperature. ARL Techn. Note 60–161, Univ. of Minnesota (1960).

[21] F. Ehlers, A. Walz: Über die Verbesserung der Profileigenschaften durch Grenzschichtbeeinflussung. Festschrift zum 60. Geburtstag von Prof. Betz (Prof. Betz's Memorial Volume), Göttingen 1945, pp. 29–35.

[22] R. Eppler: Grenzschichtberechnung mit digitalen Rechenautomaten, 3ème Congrès Aeron. Europ. 1958 II, p. 380, and Report MN 4, Bölkow-Entwicklungen K. G., Munich.

[23] R. Eppler: Praktische Berechnung laminarer und turbulenter Absauge-Grenzschichten, Ing. Arch. 32, 221–245 (1963).

[24] L. Euler: Principes généraux du mouvement des fluides. Hist. Acad. Berlin (1755) (Opera Omnia II 12, pp. 54–92).

[25] V. M. Falkner, S. W. Skan: Some approximate solutions of the boundary layer equations. Aeron. Res. Committee, Gt. Britain, Rep. 1314 (1930); Phil. Mag. 12, 865 (1931).

[26] T. Fannelöp, I. Flügge-Lotz: The laminar compressible boundary layer along a wave-shaped wall. Ing. Arch. 33, 24–35 (1963).

[27] K. O. Felsch: Beitrag zur Berechnung turbulenter Grenzschichten in zweidimensionaler inkompressibler Strömung. Thesis, Technical University Karlsruhe 1965.

[28] K. O. Felsch: Über Fortschritte in der Berechnung turbulenter Grenzschichten nach der Methode der Integralbedingungen. Mitteilungen d. Instituts für Strömungslehre und Strömungsmaschinen, Technical University Karlsruhe, Heft 4, 1967.

[29] H. Fernholz: Halbempirische Gesetze zur Berechnung turbulenter Grenzschichten nach der Methode der Integralbedingungen. Ing. Arch. 33, 384–395 (1964).

[30] H. Fernholz: Three-dimensional disturbances in a two-dimensional incompressible turbulent boundary layer. Aeron. Res. Committee, Gr. Britain. Rep. & Memorand. No. 3368, H.M. Stat. Off. London 1964.

[31] H. Fernholz: Theoretische Untersuchung zur optimalen Druckumsetzung in Unterschall-Diffusoren. Thesis, Technical University, Karlsruhe 1961.

[32] H. Fernholz: Grenzschichttheorie optimaler Diffusoren. Ing. Arch. 35, 192–201, (1966).

[33] D. Geropp: Näherungstheorie für kompressible laminare Grenzschichten mit zwei Formparametern für das Geschwindigkeitsprofil. Thesis, Technical University Karlsruhe 1963; and Deutsche Versuchsanstalt für Luftfahrt. Report No. 288.

[34] D. Geropp: Eine ähnliche Lösung der kompressiblen laminaren Grenzschichtgleichungen für eine Düsenströmung. Acta Mechanica I, No. 4, 272–281 (1965).

[35] D. Geropp: Eine ähnliche Lösung der kompressiblen laminaren Grenzschichtgleichungen für eine Düsenströmung. Z. f. angew. Math. u. Mech. 46, 195–198, Special Issue 1966.

[36] D. Geropp: Bemerkung zum System der unendlich vielen Integralbedingungen der Grenzschichttheorie. Z. f. angew. Math. u. Mech. 48, 492–493 (1968).

[37] S. Goldstein: A note on the boundary layer equations. Proc. Cambr. Phil. Soc. 35, 338 (1939).

[38] H. Görtler: Ein Differenzenverfahren zur Berechnung laminarer Grenzschichten. Ing. Arch. **16**, 173–187 (1948).

[39] H. Görtler: Zur Approximation stationärer laminarer Grenzschichtströmungen mit Hilfe der abgebrochenen Blasiusschen Reihe. Arch. d. Math. **1**, No. 3, 235 (1949).

[40] P. S. Granville: The calculation of the viscous drag of bodies of revolution. U.S. Navy Department, David Taylor Model Basin, Rep. No. 849 (1953).

[41] E. Gruschwitz: Die turbulente Reibungsschicht in ebener Strömung mit Druckabfall und Druckanstieg. Ing. Arch. **2**, 321–346 (1931).

[42] W. Hantzsche, H. Wendt: Die laminare Grenzschicht an der ebenen Platte mit und ohne Wärmeübergang unter Berücksichtigung der Kompressibilität. Jahrb. d. dt. Luftfahrtforschung **I**, 40–50 (1942).

[43] D. R. Hartree: On an equation occurring in Falkner and Skan's approximate treatment of the equations of the boundary layer. Proc. Cambr. Phil. Soc. **33**, Part II, 223 (1937).

[44] M. R. Head: An approximate method of calculating the laminar boundary layer in two-dimensional incompressible flow. Aeron. Res. Committee, Gr. Britain, Rep. and Mem. 3121 (1957).

[45] K. Hiemenz: Die Grenzschicht an einem in den gleichförmigen Flüssigkeitsstrom eingetauchten geraden Kreiszylinder. Thesis, Göttingen 1911. Dingl. Polytechn. J. **326**, 321 (1911).

[46] J. O. Hinze: Turbulence. McGraw-Hill Book Company, Inc. New York, Toronto, London 1959.

[47] H. Holstein, T. Bohlen: Ein einfaches Verfahren zur Berechnung laminarer Reibungsschichten, die dem Näherungsansatz von K. Pohlhausen genügen. Lilienthal-Bericht S 10, 5–16 (1940).

[48] L. Howarth: Concerning the effect of compressibility on laminar boundary layers and their separation. Proc. Roy. Soc. London A 194, 16 (1948).

[49] L. Howarth: On the solution of the laminar boundary layer equations. Proc. Roy. Soc. London A 164, 547 (1938).

[50] L. Howarth: On the calculation of steady flow in the boundary layer near the surface of a cylinder in a stream. Aeron. Res. Committee, Gr. Britain, Rep. 1632 (1935).

[51] B. Hudimoto: Momentum equations of the boundary layer and their applications to the turbulent boundary layer. Mem. Fac. Engineering, Kyoto Univ. **13**, 162–173 (1951).

[52] K. Jacob, F. W. Riegels: Berechnung der Druckverteilung endlich-dicker Profile ohne und mit Klappen und Vorflügeln. Z. f. Flugwissenschaften **11**, 357–367 (1963).

[53] M. Jischa: Die Berechnung laminarer dissoziierter Hyperschallgrenzschichten mit Integralbedingungen. Thesis, Technical University Berlin, 1968; and Deutsche Versuchsanstalt für Luft-und Raumfahrt Rep. No. 872 (1969).

[54] Th. von Kármán: Mechanische Ähnlichkeit und Turbulenz. J. Aeron. Sci. **1** (1934); *see also* NACA TM 611, (1931) and Collected Works **II**, 322–346.

[55] Th. von Kármán: The analogy between fluid friction and heat transfer. Trans. ASME **61**, 705–710, (1939); *see also* Collected Works **III**, 355–367.

[56] Th. von Kármán: Über laminare und turbulente Reibung. Z. f. Angew. Math. u. Mech. **1**, 233 (1912). Engl. Translation: NACA Techn. Mem. 1092.

[57] J. Kestin, P. D. Richardson: Wärmeübertragung in turbulenten Grenzschichten. Forsch. Ing.-Wes. **29** No. 4, 93–104, (1963).

[58] S. J. Kline, A. V. Lisin, B. A. Waitman: An experimental investigation of the effect of free stream turbulence on the turbulent boundary layer growth. Final report on contract NAW-6500 to NACA. Department of Mech. Engineering, Stanford University Rep. MD-2 (1958).

[59] F. Koschmieder, A. Walz: Ein neuer Ansatz für das Geschwindigkeitsprofil der laminaren Reibungsschicht. Lilienthal-Bericht 141, 8 (1941).

[60] L. S. G. Kovasznay: Turbulence in compressible and electrically conductive media. Mécanique de la Turbulence. Colloques Internationaux du Centre National de la Recherche Scientifique, Marseille 1961. Editions du CNRS, Paris (1962).

[61] Kwang-Tzu-Yang: An improved integral procedure for compressible laminar boundary analysis. J. Appl. Mech. 28, Series E, No. 1 (1961).

[62] Kwang-Tzu Yang: On an improved Kármán-Pohlhausen's integral procedure and a related error criterion. Proc. Fourth U.S. Nat. Congr. Appl. Mech., Univ. of Calif., Berkeley, June 18–21, 1962, Vol. 2, 1419–1429, New York, American Society of Mech. Engineers, 1962.

[63] Landolt-Börnstein: Zahlenwerte und Funktionen aus Physik, Chemie, Astronomie, Geophysik und Technik, Vol. IV/1. 6th ed., Springer-Verlag, Berlin, 1960.

[64] J. Laufer: Sound radiation from a turbulent boundary layer. Mécanique de la turbulence. Colloques Internationaux du Centre National de la Recherche Scientifique, Marseille 1961. Editions du CNRS, Paris (1962).

[65] L. S. Leibenson: Energy from of the integral conditions in the boundary layer theory. CAHI Rep. No. 240, 41–44 (1935).

[66] T. Y. Li, H. T. Nagamatsu: Similar solutions of compressible boundary layer equations. J. Aeron. Sc. 22, 607–616 (1955).

[67] H. Ludwieg, W. Tillmann: Untersuchungen über die Wandschubspannung in turbulenten Reibungsschichten. Ing. Arch. 17, 207–218 (1949).

[68] W. Mangler: Die "ähnlichen" Lösungen der Grenzschichtgleichungen. Z. f. angew. Math. u. Mech. 23, 243 (1943).

[69] W. Mangler: Das Impulsverfahren zur näherungsweisen Berechnung der laminaren Reibungsschicht. Z. f. angew. Math. u. Mech. 24, 251–256 (1944).

[70] W. Mangler: Zusammenhang zwischen ebenen und rotationssymmetrischen Grenzschichten in kompressiblen Medien. Z. f. angew. Math. u. Mech. 28, 97–103 (1948).

[71] M. Mayer: Theoretische und experimentelle Untersuchungen über die Zerfaserung thermoplastischer Stoffe in Blasdüsen. Thesis, Technical University Karlsruhe 1964.

[72] Mécanique de la Turbulence, Colloques Internationaux du Centre National de la Recherche Scientifique, Marseille 1961, Editions du CNRS, Paris (1962).

[73] Flyer on Material Properties, publ. by German Shell Co.

[74] M. V. Morkovin: Effects of compressibility on turbulent flows. Mécanique de la Turbulence, Colloques Internationaux du Centre National de la Recherche Scientifique, Marseille 1961, Editions du CNRS, Paris (1962).

[75] D. N. Morris, J. W. Smith: The compressible laminar boundary layer with arbitrary pressure and surface temperature gradients. J. Aeron. Sc. 20, 805–818 (1953).

[76] H. L. Moses: The Behavior of Turbulent Boundary Layers in Adverse Pressure Gradients. MIT Gas Turbine Laboratory, Rep. No. 73, Cambridge, Massachusetts, Jan. 1964.

[77] M. Navier: Mémoire sur les lois du mouvement des fluides. Mém. de l'Acad. d. Sci. 6, 389 (1827).

[78] A. Newman: Some contributions to the study of the turbulent boundary near separation. Austral. Dep. Supl. Rep. No. ACA-53 (1951).

[79] I. Newton: Philosophiae naturalis principia mathematica. Vol. 2, 1723.

[80] K. Nickel: Eine einfache Abschätzung für Grenzschichten. Ing. Arch. 31, 85–100 (1962).

[81] K. Nickel: Die Prandtlschen Grenzschicht-Differentialgleichungen als asymptotischer Grenzfall der Navier-Stokesschen und der Eulerschen Differentialgleichungen. Arch. Rat. Mech. and Analysis. 13, No. 1, 1–14 (1963).

[82] K. Nickel: Allgemeine Eigenschaften laminarer Grenzschichtströmungen. In: Grenzschichtforschung, IUTAM-Symposium Freiburg i. Br. 1957, H. Görtler, ed., Springer-Verlag, Berlin, 1958.

[83] J. Nikuradse: Laminare Reibungsschichten an der längsangeströmten Platte. Monograph, Zentrale f. wiss. Berichtswesen, Berlin 1942.

[84] O. A. Oleinik: On a system of equations in the boundary layer theory (Russian) Zh. Vychislitel'noi Mat. i Mat. Fiz. **3**, 3, 489–507 (1963).

[85] K. Oswatitsch: Gasdynamik. Springer-Verlag, Wien 1952.

[86] R. P. Patel: An improved law for the skin friction in an incompressible turbulent boundary layer in any pressure gradient. Department of Mechanical Engineering McGill Univ. Montreal, Rep. 62–4 (1962).

[87] H. H. Pearcey, D. W. Holder: Simple methods for the prediction of wing buffeting resulting from bubble type separation. 14th Meeting of the AGARD Structures and Materials Panel, Paris, July 3–6, 1962 (NPL Aero Rep. 1024).

[88] W. Pfenninger: Untersuchungen über Reibungsverminderung an Tragflügeln, insbesondere mit Hilfe von Grenzschichtabsaugung. Inst. für Aerodyn., Federal Institute of Technology, Zürich, Report No. 13 (1946). *See also* J. Aeron. Sc. **16**, 227 (1949); NACA Tech. Memo. No. 1181 (1947).

[89] N. A. V. Piercy, G. H. Preston: A simple solution of the flat plate problem of skin friction and heat transfer. Phil. Mag. (7) **21**, 996 (1936).

[90] R. M. Pinkerton: Calculated and measured pressure distributions over the midspan section of the NACA 4412 airfoil. NACA Rep. 563 (1936).

[91] K. Pohlhausen: Zur näherungsweisen Integration der Differentialgleichungen der laminaren Reibungsschicht. Z. f. angew. Math. u. Mech. **1**, 252–268 (1921).

[92] E. Pohlhausen: Der Wärmeaustausch zwischen festen Körpern und Flüssigkeiten mit kleiner Reibung und kleiner Wärmeleitung. Z. f. angew. Math. u. Mech. **1**, 115 (1921).

[93] P. P. Polentz, W. A. Page, L. L. Levy Jr.: The unsteady normal force characteristic of selected NACA profiles at high subsonic Mach numbers. NACA RM A 55C 02 (1955).

[94] L. Prandtl: Über Flüssigkeitsbewegung bei sehr kleiner Reibung. Verhandlg. d. III. Intern. Math. Kongr. Heidelberg 1904. Reprinted in: Vier Abhandlungen zur Hydrodynamik und Aerodynamik, Göttingen 1927, and NACA Tech. Memo. No. 452 (1928).

[95] L. Prandtl: Führer durch die Strömungslehre (Chapter on Turbulence). 4th ed., Verlag Fr. Vieweg & Sohn, Braunschweig 1956.

[96] L. Prandtl in: Aerodynamic Theory; William F. Durand, Editor, Vol. III. Springer, Berlin 1935.

[97] L. Prandtl, H. Schlichting: Das Widerstandsgesetz rauher Platten. Werft, Reederei, Hafen 1–4 (1934).

[98] J. Pretsch: Die Stabilität einer ebenen Laminarströmung bei Druckgefälle und Druckanstieg. Jahrb. d. dt. Luftfahrtforschung **1**, 58 (1941).

[99] J. Pretsch: Die Anfachung instabiler Störungen in einer laminaren Reibungsschicht. Jahrb. d. dt. Luftfahrtforschung **1**, 54–71 (1942).

[100] J. Pretsch: Zur theoretischen Berechnung des Profilwiderstandes, Jahrb. d. dt. Luftfahrtforschung **1**, 61 (1938). Engl. Translation, NACA Tech. Memo. No. 1009 (1942).

[101] A. W. Quick, K. Schröder: Verhalten der laminaren Grenzschicht bei periodisch schwankendem Druckverlauf. Math. Nachr. **8**, 217–238 (1953).

[102] Lord Rayleigh: On the stability of certain fluid motions. Proc. London Math. Soc. **11**, 57 (1880) and **19**, 67 (1887); Scientific Papers **I**, 474 (1880); **III**, 17 (1887); **IV**, 203 (1895); **VI**, 197 (1913).

[103] J. Rechenberg: Zur Messung der turbulenten Wandreibung mit dem Preston-Rohr. Jahrb. d. Wiss. Ges. f. Luft-u. Raumfahrt 151–159 (1962).

[104] J. Rechenberg: Messung der turbulenten Wandschubspannung. Z. f. Flug-
wissenschaften **11**, 429–438 (1963).

[105] H. Reichardt: Vollständige Darstellung der turbulenten Geschwindigkeitsver-
teilung in glatten Leitungen. Z. f. angew. Math. u. Mech. **31**, 208–219 (1951).

[106] O. Reynolds: (a) An experimental investigation of the circumstances which determine whether the motion of water shall be direct or sinuous, and of the law of
resistance in parallel channels. *Phil. Trans. Roy. Soc. Lond.*, vol. 174, pt.
3, 935–
982, (1883); (b) On the dynamical theory of incompressible viscous fluids. *Ibid.*,
vol. 186A, 123–164, (1895).

[107] J. Rotta: Schubspannungsverteilung und Energie-Dissipation bei turbulenten
Grenzschichten. Ing. Arch. **20**, 195–206 (1952).

[108] J. Rotta: Turbulent boundary layers in incompressible flow. In: Progress in Aero-
nautical Sciences, Vol. II, Pergamon Press, London 1962.

[109] J. Rotta, K. G. Winter, K. G. Smith: Untersuchungen der turbulenten Grenzschicht an einem taillierten Drehkörper bei Unter- und Überschallströmung.
Aerodyn. Versuchsanstalt Report No. 65 A 21 a (1965).

[110] R. Sauer: Einführung in die theoretische Gasdynamik. Springer-Verlag, Berlin,
Göttingen, Heidelberg 1960.

[111] H. Schlichting: Boundary Layer Theory. Transl. from the German by J. Kestin,
McGraw-Hill New York, 1960, 4th ed. [see also new 6th ed., 1968].

[112] H. Schlichting: Über die theoretische Berechnung der kritischen Reynolds-Zahl
einer Reibungsschicht in beschleunigter und verzögerter Strömung. Jahrb. d. dt.
Luftfahrtforschung **1**, 97 (1940).

[113] H. Schlichting: Zur Entstehung der Turbulenz bei der Plattenströmung, Nachr.
Ges. Wiss. Göttingen, Math.-Phys. Kl. 182–208 (1933); *see also* Z. f. angew.
Math. u. Mech. **13**, 171 (1933).

[114] H. Schlichting: Der Wärmeübergang an einer längsangeströmten Platte mit
veränderlicher Wandtemperatur. Forsch. Ing.-Wes. **17**, 1 (1951).

[115] K. E. Schoenherr: Resistance of flat surfaces moving through a fluid. Trans. Soc.
Nav. Arch. & Mar. Eng. **40**, 279, (1932).

[116] W. Schönauer: Die Lösung der Croccoschen Grenzschichtgleichung. Thesis,
Technical University Karlsruhe 1963.

[117] O. Schrenk, A. Walz: Theoretische Verfahren zur Berechnung von Druck- und
Geschwindigkeitsverteilungen, Jahrb. d. dt. Luftfahrtforschung **I**, 29 (1939).

[118] G. B. Schubauer, P. S. Klebanoff: Investigation of separation of the turbulent
boundary layer. NACA Rep. 1030 (1951).

[119] G. B. Schubauer, H. K. Skramstadt: Laminar boundary layer oscillations and
stability of laminar flow. Nat. Bureau of Standards Research Paper 1772 (1943).
see also J. Aeron. Sc. **14**, 69(1947); and NACA Rep. 909 (1948), and NACA ACR
(1943).

[120] G. B. Schubauer, W. G. Spangenberg: Forced mixing in boundary layers. J.
Fluid Mech. **8**, 10–31 (1960).

[121] F. Schultz-Grunow: Neues Widerstandsgesetz für glatte Platten. Luftfahrtforschung **17**, 239 (1940); *see also* NACA T. M. No. 986 (1941).

[122] S. F. Shen: Calculated amplified oscillations in plane Poiseuille and Blasius flows.
J. Aeron. Sc. **21**, 62–64 (1954).

[123] D. B. Spalding: Heat transfer to a turbulent stream from a surface with a step-
wise discontinuity in wall temperature. Special issue on International Develop-
ments in Heat Transfer. Published by Amer. Soc. Mech. Engrs. 439–446 (1961).

[124] D. B. Spalding: A new procedure for the numerical solution of the elliptic equations of simultaneous heat, mass, and momentum transfer. AGARD Seminar on
Numerical Methods of Viscous Flows, Teddington, Sept. 18–21, 1967.

[125] H. Sprenger: Experimentelle Untersuchungen an geraden und gekrümmten
Diffusoren. Thesis, Federal Institute of Technology Zürich, 1959.

[126] H. B. Squire, A. D. Young: The calculation of the profile drag of airflow. Aeron. Res. Committee, Gr. Britain, Rep. 1838 (1938).

[127] G. G. Stokes: On the theories of the internal friction of fluids in motion. Trans. Cambr. Phil. Soc. **8**, 227–319 (1849).

[128] B. S. Stratford: The prediction of separation of a turbulent layer. J. Fluid Mech. **5**, 1–16 (1959).

[129] B. S. Stratford: An experimental flow with zero skin friction throughout its region of pressure rise. J. Fluid Mech. **5**, 17–35 (1959).

[130] G. Streit, F. Thomas: Experimentelle und theoretische Untersuchungen an Ausblaseflügeln und ihre Anwendung beim Flugzeugentwurf. Jahrb. d. Wiss. Ges. f. Luft u. Raumfahrt, 119–132 (1962).

[131] D. M. Sutherland: The viscosity of gases and molecular force. Phil. Mag. **36**, 507 (1893).

[132] J. Tani: On the solution of the laminar boundary layer equations, in: 50 Jahre Grenzschichtforschung. Vieweg Verlag, Braunschweig 1955.

[133] T. Theodorsen, J. Garrick: General potential theory of arbitrary wing sections. NACA Rep. 452 (1933).

[134] F. Thomas: Ermittlung der Buffeting-Grenzen von Tragflügeln. Report 65/6 of the Dornier-Werke Friedrichshafen (1965).

[135] B. G. J. Thompson: A critical review of existing methods of calculating the turbulent boundary layer. Aeron. Res. Committee, Gr. Britain, 26 109, F. M. 3492 (1964).

[136] W. Tollmien: Über die Entstehung der Turbulenz, 1. Mitteilung Nachr. Ges. Wiss. Göttingen, Math.-Phys. Klasse, 21–44 (1929). For additional papers by Tollmien on stability, see [111], p. 408.

[137] A. A. Townsend: The development of turbulent boundary layers with negligible wall stress. J. Fluid Mech. **8**, 143–155 (1960).

[138] E. Truckenbrodt: Ein Quadratur-Verfahren zur Berechnung der laminaren und turbulenten Reibungsschicht bei ebener und rotationssymmetrischer Strömung. Ing. Arch. **20**, 211–228 (1952).

[139] VDI-Wärmeatlas, Deutscher Ing. Verlag GmbH, Düsseldorf (1954).

[140] A. Walz: Anwendung des Energiesatzes von Wieghardt auf einparametrige Geschwindigkeitsprofile in laminaren Grenzschichten. Ing. Arch. **16**, 243–248 (1948).

[141] A. Walz: Beitrag zur Näherungstheorie kompressibler turbulenter Grenzschichten. Deutsche Versuchsanstalt für Luftfahrt Repts. No. 84 and 136 (1954, 1960).

[142] A. Walz: Beitrag zur Näherungstheorie kompressibler laminarer Grenzschichten mit Wärmeübergang. Deutsche Versuchsanstalt für Luftfahrt Rep. No. 281 (1963).

[143] A. Walz: Nouvelle méthode approchée de calcul des couches limites laminaires et turbulentes en écoulement compressible. Publ. Sci. Techn. du Min. de l'Air No. 309 and 336 (1956, 1957).

[144] A. Walz: Zur Anwendung des Druckverteilungs-Rechenverfahrens von Theodorsen-Pinkerton. Aerodyn. Versuchsanstalt Forschungsbericht UM No. 3160 (1944).

[145] A. Walz: Zur Berechnung der Druckverteilung an Klappenprofilen mit Totwasser. Jahrb. d. dt. Luftfahrtforschung 1940.

[146] A. Walz: Theorie zur Absaugung der Reibungsschicht, Zeitschr. Wissenschaftl. Berichtswesen Report 1775, Aerodyn. Versuchsanstalt Göttingen (1943) (available from Zentralstelle f. Luftf.-Dokum. u. Inform. München, Airport.)

[147] A. Walz: Näherungstheorie zur Grenzschichtabsaugung durch Einzelschlitze. Deutsche Versuchsanstalt für Luftfahrt Rep. No. 184 (1962).

[148] A. Walz: Einige Beispiele zur theoretischen Berechnung der Polaren eines Tragflügelprofils. Forsch.-Ber. d. Z. f. wiss. Ber.-Wesen, Rep. No. 1848 (1943).

[149] A. Walz: Zur theoretischen Berechnung des Höchstauftriebsbeiwertes von Trag-flügelprofilen mit und ohne Auftriebsklappen. Forsch.-Ber. d. Z. f. wiss. Ber.-Wesen Rep. No. 1769 (1943).

[150] A. Walz: Näherungsverfahren zur Berechnung der laminaren und turbulenten Reibungsschicht (Appendix: Determination of $c_{a\,max}$- and c_{wp}). Untersuchungen und Mitteilungen der Aerodyn. Versuchsanstalt Göttingnen UM No. 3060(1943).

[151] A. Walz: Über Fortschritte in Näherungstheorie und Praxis der Berechnung kompressibler laminarer und Grenzschichten mit Wärmeübergang. Z. f. Flug-wissenschaften 13, 90–102 (1965).

[152] A. Walz: Graphische Hilfsmittel zur Berechnung der laminaren und turbulenten Reibungsschicht. Thesis, Technical University Braunschweig 1941, Report S 10 Lilienthal-Gesellschaft, 45–74.

[153] A. Walz: Neue Anwendung des Prinzips der gemittelten Grenzschichtbedingungen nach v. Kármán und Pohlhausen. Grenzschichtforschung (IUTAM Symposium, Freiburg 1957), H. Görtler, ed., Springer Verlag, Berlin, 1958.

[154] H. Weyl: Concerning the differential equations of some boundary layer problems. Proc. Nat. Ac. Sci., Washington 27, 578 (1941).

[155] K. Wieghardt: Über einen Energiesatz zur Berechnung laminarer Grenzschichten. Ing. Arch. 16, 231 (1948) and internal report, Kaiser-Wilhelm Institut, Göttingen (1944).

[156] K. Wieghardt: Friction drag on rough plates. Z. f. angew. Math. u. Mech. 24, (1944).

[157] H. W. Wipperman: Ein Differenzenverfahren zur Lösung der Stewartson-Crocco-Gleichungen für kompressible laminare Grenzschichten. Thesis, Technical University Karlsruhe 1966.

[158] H. Witting: Verbesserung des Differenzenverfahrens von H. Görtler zur Berechnung laminarer kompressibler Grenzschichten. Thesis Univ. Freiburg i. Br. 1961.

[160] J. Zierep: Vorlesungen über theoretische Gasdynamik. Verlag G. Braun, Karlsruhe 1963.

Subject Index

Name Index